Participatory Monitoring and Evaluation

About the Editor

Dr. Maravalalu J. Chandre Gowda (b. 1965) is currently working as Additional Commissioner (Extn) in the Department of Agriculture and Cooperation, Ministry of Agriculture, Government of India. He has served 20 years in various organisations such as Indian Institute of Horticultural Research, Zonal Coordinating Unit-VIII of ICAR Transfer of Technology Projects and Department of Agriculture and Cooperation, Ministry of Agriculture, Government of India. His international experiences include training at Asian Vegetable Research and Development Centre (AVRDC), Taiwan, and participating in FAO Consultation Workshop on Market Oriented Agricultural Extension at Bangkok, Thailand. Dr. Gowda has rich experience of guiding the national agricultural extension programmes like Extension Reforms (ATMA), Agric Clinics and Agri Business Centres, Kisan Call Centres, Model Training Courses and Extension Education Institutes during his tenure in Ministry of Agriculture.

Dr. Chandre Gowda has edited 6 books, published 25 research papers in refereed journals and has equal number of symposia papers. He is the recipient of Young Scientist award of the Indian Society of Extension Education. He has been honoured with **Utkrisht Krishi Seva Trophy** by the Ministry of Agriclture, Government of India for his exemplary work in the field of Advancement of Agriculture in the country.

Participatory Monitoring and Evaluation

Edited by

M.J. Chandre Gowda

CONCEPT PUBLISHING COMPANY PVT. LTD.,
NEW DELHI-110059

ISBN-13 : 978-81-8069-831-6

First Published 2012

Published and Printed by

Concept
Since 1974

Concept Publishing Company Pvt. Ltd.
A/15-16, Commercial Block, Mohan Garden
New Delhi-110059 (India)
T: 25351460, 25351794, *F*: 091-11-25357109
E: publishing@conceptpub.com, W: www.conceptpub.com
Editorial Office: H-13, Bali Nagar, New Delhi-110 015, India.

Cataloging in Publication Data--*Courtesy:* D.K. Agencies (P) Ltd. <docinfo@dkagencies.com>

Participatory monitoring and evaluation / edited by M.J. Chandre Gowda.
p. cm.
Includes bibliographical references and index.
ISBN 9788180698316

1. Participatory monitoring and evaluation (Project management)—Congresses. 2. Participatory monitoring and evaluation (Project management)—India—Congresses. 3. Agricultural extension work—Evaluation—Congresses. 4. Agricultural extension work—India—Evaluation—Congresses. 5. Participatory rural appraisal—Congresses. 6. Participatory rural appraisal—India—Congresses. I. Gowda, M. J. Chandre (Maravalalu Jauaregowda Chandre), 1965-

DDC 658.404 23

Preface

Monitoring and Evaluation are the two integrated tools for ascertaining the efficiency and effectiveness of investments in any area. Though the immediate focus of any Monitoring and Evaluation process is on intermediate outcomes, the larger focus is on a learning approach that uses achievements and problems for better decision-making and accountability. Monitoring is often seen as an obligation imposed by the funding agencies. The lack of commitment among staff towards monitoring leads to little use of information gathered during monitoring. Information gathered in situations like this remains as data collected for writing reports, thereby ignoring factors like project outreach, effect and impact.

Much has been debated on how to make monitoring and evaluation really interesting to the practitioners and useful to all stakeholders in terms of the end result. Consequently, increased attention is now being paid to monitor the needs of other stakeholders, such as beneficiaries, community-based organizations and other local partners. Literature available on Monitoring and Evaluation (M&E) appears to be inadequate, quantitatively as well as qualitatively, to propagate Participatory M&E of varied nature of programmes and schemes. Documentation of methodologies and processes in a way that promotes adoption by the practitioners at the operational level and by the trainers for HRD purposes are required in huge numbers as the users and the conditions under which they use M&E tools are infinite. The present publication is an attempt to collate useful experiences related to Participatory Monitoring and Evaluation of agriculture and rural development programmes. Its uniqueness lies in the balanced presentation of theoretical and practical aspects and also a right mix of conventional and participatory approaches to monitoring and evaluation.

Editor

Acknowledgements

This is a compilation of the long and varied experiences of agricultural extension scientists and professionals in the planning and implemention of agriculture and rural development programmes. Most of these were the discussion papers written by the authors for interaction with the participants of the winter school on Participatory Impact Monitoring and Assessment sponsored by the Indian Council of Agricultural Research. These interactions were so absorbing and effective that one of the participants, as part of feedback on the course during the valediction, rated the proceedings as 'international level', and the learnings as 'down to earth'. The discussion papers were further developed and have been compiled in this book.

Professor H.N. Byra Reddy deserves the highest level of gratitude for an in-depth delving into the fundamentals of monitoring and evaluation. Whole-hearted appreciations go to Dr. N. Nagaraja for sharing his rich experiences on the innovative approaches to participatory monitoring and evaluation. Thanks to Ms. Vidya Ramachandran for bringing the participatory impact assessment to the core of monitoring and evaluation processes.

Contents

List of Contributors

Adapur, S.H. (Mr.), Subject Matter Specialist (Agril. Extension), KVK Hulukoti, Gadag, Karnataka, India.

Adhiguru, P. (Dr.), Senior Scientist (Agril. Extension), National Centre for Agricultural Economics and Policy Research (NCAP), P.B. No. 11305, DPS Marg, Pusa, New Delhi-110 012, India.

Byra Reddy, H.N. (Dr.), Former Director of Extension, Andhra Pradesh Agricultural University, No. 55C, 7th Cross, UAS Layout, Bangalore-560094, Karnataka, India.

Chandre Gowda, M.J. (Dr.), Additional Commissioner (Extension), Room No. 527A, Krishi Bhavan, Department of Agriculture and Cooperation, Ministry of Agriculture, Government of India, New Delhi, 110001

Nagaraja, N. (Dr.), Director of Instruction, Sericulture College, Chintamani, Karnataka, India.

Narayana, Gowda K. (Dr.), Project Coordinator, Regional Bio Resource Centre and, Registrar, University of Agricultural Sciences, Bangalore-560065, Karnataka, India.

Manjula, N. (Dr.), Subject Matter Specialist (Agril. Extension), KVK Chintamani, Karnataka, India.

Palani Vel Murugan, C. (Mr.), Research Fellow, National Centre for Agricultural Economics and Policy Research (NCAP), P.B. No. 11305, DPS Marg, Pusa, New Delhi-110012, India.

Ramachandran, Vidya (Ms.), 371, 100 Ft. Road, HAL II Stage, Bangalore- 560008, Karnataka, India.

Savitha, C.M. (Ms.), Research Associate, Community Based Tank Management Project Consultancy Services, University of

Agricultural Sciences, GKVK, Bangalore-560065, Karnataka, India.

Varadaraju, G.M. (Dr.), Associate Professor of Extension, Directorate of Extension, University of Agricultural Sciences, Hebbal, Bangalore, 560024, Karnataka, India.

Introduction

Monitoring and Evaluation (M&E) are the two closely associated terms in any project implementation process. Budget, duration, agencies of funding, the implementing agencies, the beneficiaries, and the other stakeholders have played varying roles in the nature and intensity of monitoring and evaluation carried out in a project. Nevertheless, they have remained an integral part of any project cycle. Monitoring is often referred to as a regular activity for keeping track of what is happening and in the process, and documenting what changes are taking place. Evaluation, by contrast, takes note of those changes to assess whether these changes are desirable and adequate. Proper monitoring and evaluation address all aspects of a project including the process and product dimensions of a project. Monitoring and evaluation are inter-dependent for success of each other. Evaluation should not take place without the inputs and data generated during monitoring. Therefore, evaluation must be a concurrent and on-going process and happens along with monitoring and not at the end of the project cycle alone, which happens quite often. Since monitoring generates huge amount of information on a continuous basis, its interpretation (which is the role of evaluation) must also happen on a continuous basis. Similarly, monitoring, oblivious of the evaluation process and its findings may not contribute in any way for the betterment of the project. It is thus imperative that monitoring and evaluation co-exist and must complement and supplement each other.

Participation is increasingly being recognized as an integral part of the M&E process, since it offers new ways of assessing and learning from change that are more inclusive and more responsive to the needs and aspirations of those most directly affected. Participatory Monitoring and Evaluation (PM&E) is geared towards not only measuring the effectiveness of a project, but also towards building ownership and empowering beneficiaries; building accountability and transparency; and taking corrective actions to improve performance and outcomes.

Monitoring is a periodic but regular activity for 'keeping track' of what is happening in any project intervention. It stands for knowing where we are through the process of observing and recording changes at regular intervals. Evaluation, by contrast, happens at a pre-defined point within a long period of project interventions and entails a reflection and assessment of what has been achieved and learned.

The system of monitoring and evaluation should form the core of project design itself. It should start before the commencement of project activities, with decisions on what should be monitored, and eventually evaluated. These criteria are often defined as indicators. Furthermore, if the system of monitoring and evaluation is to be truly participatory, the indicators and the means of determining them should be discussed, identified and agreed by the community stakeholders at the beginning. Same stakeholders must decide how often progress should be reviewed, who should do it and what methods be used. The word 'Participatory' implies shared learning, joint decision-making, co-ownership, democratic process to involve everyone in the community, mutual respect, empowerment and enhanced mutual understanding.

Participatory Monitoring and Evaluation enables the stakeholders to engage in monitoring and evaluating, exchange the contents, the processes and the results of the M&E activity and take up corrective actions. It is not just a matter of using participatory techniques within a conventional monitoring and evaluation setting. It is about radically rethinking who initiates and undertakes the process and who learns or benefits from the findings (Estrella and Gaventa, 1998).

The four principles of PM&E are, (i) 'Participation' which means opening up the design of the M&E system to include those most directly affected and agreeing to analyze data together, (ii) 'Negotiation' among the programme managers, implementers and community members to agree on what will be monitored or evaluated, how and when the data will be collected and analyzed, what the data actually means, how the findings will be shared and what actions will be taken, (iii) 'Learning' by all those concerned from the process and contributions of other stakeholders, and (iv) 'Flexibility' to enable number, role and skills of stakeholders and other factors change over time (Judi, 2004).

Aims of Participatory Monitoring and Evaluation (Jain and Wim Polman, 2003) are: (i) to assess information or generate data on development activities being carried out at the local community level, (ii) to facilitate monitoring and evaluation by beneficiaries of different development activities; and (iii) to increase beneficiaries' commitment and understanding in designing, planning and implementing community-based development projects or programmes.

Core Steps in Participatory Monitoring and Evaluation (Irene, 2000)

1. Identify who should, and wants to be involved.
2. Clarify participants' expectations of the process (what are their information needs), and in what way each person or group wants to contribute.
3. Define the priorities for monitoring and evaluating (on which goals/objectives/activities to focus).
4. Identify indicators that will provide the information needed.
5. Agree on the methods, responsibilities and timing of information collection.
6. Collect the information.
7. Adapt the data collection methodology, as needed.
8. Analyze the information.
9. Agree on how the findings are to be used and by whom.
10. Clarify if the PM&E process needs to be sustained, and if so, how. Adjust the methodology accordingly.

The book is a confluence of concepts, approaches, and case studies related to the world of monitoring, evaluation and impact assessment. More emphasis is given to Impact Assessment in this publication, not only to contest the least importance given to it hitherto, but also to press for its importance in understanding why programmes do not reach the people for whom these programmes are meant and fail to achieve the changes aimed at the planning and designing stage. It is also required to know the significance of the changes that have happened amidst the quagmire development scenario projected in most situations.

The chapter "Monitoring, Evaluation and Impact Assessment" describes the concept of evaluation in great detail. Enumeration of types and stages of evaluation is followed by a brief clarification on levels and degrees of evaluation. Important considerations for impact assessment are credibility, usefulness and cost effectiveness of monitoring and evaluation, which is further facilitated by carefully selected set of indicators. *Indicators are quantitative as well as qualitative variables that can be measured or described and when carried out periodically, have the potential to provide all that is required for monitoring and evaluation.* Indicators of changes in process, performance and impact of the project must provide a direct and unambiguous measure of the progress. Indicators must be sensitive enough to reflect the minor changes but at the same time must not easily be blown of course by unrelated developments. It is better achieved through a participatory process, by involving all the stakeholders—all those involved and affected by the project.

Monitoring and evaluation is carried out by employing appropriate M&E tools, methods and designs. The choice of tools varies according to the methods (qualitative and quantitative) of evolution. The choice of methods depends on the purpose and its audience. Designs are essential to avoid threats to internal and external validity of the evaluation process. Well designed evaluation enables the evaluator to say with confidence that which impacts are due to the project and which are influenced by outside factors. Three kinds of designs (pre-experimental, experimental and quasi-experimental) are used by the evaluating agencies, but the most commonly used is the pre-experimental design, which focuses on a single programme and attempt to discover whether or not the project has achieved its objectives. Though pre-experimental design is simple to use, evaluators cannot be confident if changes were caused by the project intervention or by a host of external factors.

Qualitative methods address questions related to 'how and why' of the project implementation and are best at uncovering reasons behind unanticipated changes. On the other, quantitative methods address questions related to what, who, where, how many, how much and the influence or prevalence of a phenomenon. Both qualitative and quantitative methods have to be used in appropriate combination and in appropriate sequence to derive maximum data.

Chapter on Measurement of Social Impacts defines 'Social Impact' and then elucidates on why social impacts have to be measured and who should be involved in the process. The criteria to be followed while identifying the indicators of social impacts are also briefly stated. Techniques of impact assessment revolve around the participatory apparatus supplemented by the details on advantages and limitations of using participatory approaches.

Chapter on 'Participatory Poverty Assessment (PPA)' delves at length on the widespread anomalies in the poverty impact measurement methodologies and the significance of the new approach introduced by the World Bank, which uses both quantitative and qualitative methods. The quantitative approaches, widely used in poverty impact monitoring, have their own limitations of having a partial view of poverty, which is very complex, diverse, and area-specific realities in which the poor lives. As a result, qualitative approaches are gaining significance in poverty assessment. An illustrative list of the PRA methods and poverty-related indicators would be very useful to the practitioners of poverty assessment. Different stages of the proposed poverty assessment process like problem identification, determination of sample size, nature of sample and location of sample households, data collection, data analysis and final results give detailed account of the process itself. The three case studies illustrated to substantiate the usefulness of the participatory approaches in assessing poverty provide a veritable depth to the assessment methods and tools.

The chapter containing a case study on impact evaluation using participatory tools like matrix analysis, time-line analysis and spider web diagrams reiterates the simple yet tremendous utility of such tools for both ordinary audiences and first-time evaluators. The ripples of change, set in motion by the interventions of a KVK on a self-help group, were unearthed in an unassuming manner through a combination of time-line and matrix rating. The tools succeeded in revealing the enhanced livelihood options, increased employment opportunities, reduced dependency on private sources of credit, improved nutrition and health and improved self-respect for the group members from within and outside the family. Equally mesmerizing was the way the group members drew the spider web diagram to visually communicate the important dimensions of changes in their economic (ability to spend in emergency, household articles/ productive assets possession) and social life (status within and outside family, individual and group recognition) using locally available material.

The Participatory Impact Monitoring and Assessment (PIMA) elaborates the need for emphasizing impact assessment and the same to be done on a participatory basis. Despite the increasing number and sophistication of management tools and methodologies, monitoring the impacts of development efforts continues to be a complex and neglected task. The possible reasons for such a neglect, the operational definition of PIMA, its objectives, and clearly identified set of pre-conditions for practicing PIMA may help understand PIMA holistically. The care to be exercised while identifying the indicators for impact assessment, with appropriate examples, provides enough safeguards for a beginner. Attributing the changes and the types of changes—intended, unintended and unexpected—to the project interventions has been aptly explained. The illustration on how unexpected impacts could be both positive and negative is quite interesting. The steps involved in PIMA are bifurcated into prerequisite steps and implementation steps. The implementation process has been very clearly identified with twelve steps which are not overlapping but closely inter-dependent. Questions to be asked in each step like, the rationale, expected outputs, actors involved, procedure to be followed and cross-checking questions, will make the PIMA truly participatory.

The Logical Framework Approach (LFA) and the different nomenclatures in different countries prove its widespread acceptance. The difference between log frame and Logical Framework Approach and the details on the basic principles—a concise and a stand-alone log frame designed with the participation of beneficiaries and its use in monitoring and evaluation—set the tone for further elaboration of the approach. The four steps of a LFA, viz., situation analysis, strategy analysis, project planning matrix and the implementation have been succinctly highlighted. The project planning matrix has been further elaborated through four items—narrative summary, objectively verifiable indicators, means of verification and important assumptions. There are more advantages than limitations for LFA and thus makes it a powerful set of tools for designing project and project evaluations.

The chapter on Farmers Field School (FFS), an established participatory method for effective learning, explains FFS as a non-formal and learner-centred educational process. The key features of FFS are learning by doing, facilitation rather than teaching, interactive among the farmers as well as with facilitators, flexible but within the framework of crop situation and being unique from one another. The actors in any FFS are basically the participants, a collaborator and facilitator/s. Most important among three is the role played by the collaborator who is actually the FFS teacher. The steps involved in FFS encompass planning, implementation and post-implementation process of FFS. The planning process which involves problem analysis, development of treatment plans, designing a lay- out, working out parameters for data recording, scheduling of visits and preparation of calendar of operations is very crucial for the success of a FFS. The implementation process involves the execution of the steps highlighted in the planning process as per the calendar specified for each step. The post-implementation process centres on the activities to be carried out for withdrawal from the FFS.

Participatory Technology Development (PTD) has been a popular phrase for the Research and Development institutions in agriculture and related fields. However, not many studies have clearly outlined the processes involved and the net gains in terms of quality of the technology generated and the impacts created by such technologies. The process of technology development is rarely linked with process of social change. PTD provides an insight into linking technology development with social change. The characteristics of PTD, partners involved, the temporal and spatial dimensions of PTD processes have been explained before dealing with the stages of PTD. The stages in PTD, though theoretical in nature, have been made useful to the practitioners by actually highlighting how each step has been employed in the Community Based Tank Management Project in Karnataka. Each of the six steps of PTD has enough details to encourage even the inexperienced PTD researchers to practice it with fair degree of confidence. It provides an opportunity for blending the capabilities of scientists and the priorities of the farming communities in developing suitable technologies to the problems observed.

The chapter on grounded theory tries to score a point for the less-used, but highly potential theory-based evaluation, particularly for the people involved in programme implementation but are perplexed to know what is or is not working and why. It not only distinguishes between substantive and formal theories but also tells the readers the way the grounded theory benefits from both the theories. At least four highly interrelated properties are required for the practical application of grounded theory. They are: (i) theory must closely fit the substantive area in which it will be used, (ii) readily understandable for the layman, (iii) sufficiently general to be applicable to a multitude of diverse daily situations in the substantive area, and (iv) user's partial control over the structure and process of daily situations. The listed prerequisites for discovering grounded theories remind the seriousness and sincerity demanded for using the approach.

The book provides a complete package for monitoring and evaluation, covering the concepts, tools and methods besides case studies to demonstrate the practicing of monitoring and evaluation in everyday scenario. Assessing the impacts should be the preferred aim of any monitoring and evaluation process and hence all M&E activities should have a well-structured road map for achieving the same. Impacts cannot be measured, if not planned for achieving the same right from the planning stage. Even if it is well-planned, it may not be achieved if appropriate tools and methods are not selected; its practitioners are not adequately trained; if adequate patience is not exercised; and if the evaluators are not prepared to accept the unexpected changes due to the programmes. Let us make a new beginning in the Monitoring, Evaluation and Impact Assessment in agriculture and rural development.

References

Estrella, M. and Gaventa, J. (1998). Who counts reality? Participatory monitoring and evaluation: a literature review. Prepared for the "International workshop on participatory monitoring and evaluation: *Experiences and Lessons*" held at the *International Institute for Rural Reconstruction* (IIRR) campus, Cavite, Philippines, November 24-27, 1997. IDS Working Paper 70. Sussex, UK: Institute of Development Studies. p. 76

Irene, G., 2000. Methodological issues of participatory monitoring and evaluation. In: Learning from Change: Issues and Experiences in *Participatory Monitoring and Evaluation*. Edited by Marisol Estrella, with Jutta Blauert, Dindo Campilan, John Gaventa, Julian Gonsalves, Irene Guijt, Deb Johnson, and Roger Ricafort. IDRC/Practical Action Publishing

Jain, S.P. and W. Polman, 2003. A Handbook for Trainers on Participatory Local Development, RAP Publication 2003/07, The *Panchayati Raj Model in India*. FAO Regional Office for Asia and the Pacific, Bangkok, Thailand, August 2003.

Judi, A., 2004. Strategic Report 9, *Participatory Monitoring and Evaluation for Hygiene Improvement*. U.S. Agency for International Development, Washington, DC.

1

Monitoring, Evaluation and Impact Assessment

H.N. Byra Reddy

Introduction

Evaluation is a critical component in the development and growth of any organization, but it is especially important in the field of Agricultural Extension. A key component of evaluation is impact assessment—assessing the impact of an intervention (programme) on the clientele and the system in which it operates. Evaluation and impact assessment are essential components to improve the operational efficiency of any organization in implementing development programmes. When conducting evaluation and impact assessment, organizations want to know if their activities are furthering their mission and if they are operating in the most efficient and effective way—for clients, staff, and for donors.

Evaluation

Evaluation is a participatory process designed to determine how well a programme or project has accomplished its goals. Evaluation is always based on the examination of some established, empirical variable or indicator, and how current practices compare to that standard. The results of evaluation provide managers with information about whether to expand a programme, to continue a programme at its current level, to reduce spending, or to cut it entirely. The term "evaluation" describes different models that suit different purposes at different stages in the life of a project. Outside consultants are often hired to conduct a formal programme evaluation of an organization. As a result, evaluation has frequently been viewed as an external imposition—a process that is not very helpful to project staff. Programme staff can also conduct an internal evaluation, however. When conducted appropriately, evaluation should be a tool that not only measures success of programme, but also can contribute to it, as well.

Why Conduct an Evaluation?

Evaluation is an essential component in the development, maintenance, and performance of an organization. It helps to ensure whether a development organization is meeting its service mission and to demonstrate measurable outcomes to donors. Evaluation, therefore, is a reflective process requiring a critical look at business processes and organizational activities. Regular evaluation measures progress toward a specific goal and is a vital component of any effort to manage for results. When an organization is established to implement a project or programme, the organization determines, through strategic planning, what the mission and goals are for the organization, and the framework that will be used to implement them. This framework needs to be tested; which ensures that the organization is performing as planned, or if the organization needs to re-evaluate its processes.

Types of Evaluation

There are two basic types of evaluation : Formative and Summative.

Formative Evaluation

Formative evaluation is a tool used from the beginning to the end of a project. Typically, a formative evaluation is conducted at several points in the cycle of a project and is used to continually "form" or modify the project to make sure that its programme activities match programme goals and the overall mission. A summative evaluation assesses the project's success or failure. This type of evaluation takes place after the project is up and running, in order to judge its impact. Impact assessment can be considered synonymous with summative evaluation.

Formative evaluation is used to assess ongoing project activities. For development organizations, formative evaluation begins at the start of a project and continues throughout the life of the project. In general, formative evaluation consists of two segments: *implementation evaluation* and *progress evaluation.* Implementation evaluation collects information to determine if the programme or project is being delivered as planned. The following questions can help guide an implementation evaluation:

- Do the activities and strategies match those described in the proposal? If not, are the changes to the proposal justifiable?
- Were the appropriate staff members hired, trained, and are they working in accordance with the proposed plan?
- Were the appropriate materials and equipment obtained?
- Were activities conducted according to the proposed timeline?
- Did appropriate personnel conduct those activities?
- Were the appropriate participants selected and involved in the activities?
- Was a management plan developed and followed?

Project staff should use implementation evaluation as an internal check to see if all the essential elements of the project are in place and operating. The other aspect of formative evaluation is progress evaluation. This type of evaluation is used to assess progress in meeting the project's goals. Progress evaluation should be thought of as an interim outcome measurement. Typically, a progress evaluation will measure a series of indicators that are designed to show progress towards programme goals. These indicators for example, could include participant ratings of training, a seminar or services provided through an extension/development organization. Opinions and attitudes of participants and staff, as well as key financial indicators may also form a part of this assessment. By analyzing interim outcomes, project staff eliminates the risk of waiting until participants have experienced the entire treatment to assess outcomes. The results of an evaluation can be used broadly in an organization. The results are not only a good source of ideas for organizational improvement, but also a source of information for the organization's stakeholders, such as the Board of Directors, donors, host government, collaborators, clients or shareholders.

*2. **Summative Evaluation or Impact Assessment***

Summative evaluation or Impact assessment can be used as "management tools for aiding practitioners to better attain programme goals." In addition to the benefits that an impact assessment offers a development organization, they have an obligation to ensure that they are making the right choices in terms of their objectives. Development organizations are also accountable to their funding agencies—usually governments, shareholders and taxpayers and, therefore, they have a strong interest in obtaining measures of the effectiveness of their funds. Donors may use impact assessment results to make resource allocation decisions for individual organizations, or for broad strategic funding decisions. Summative evaluation is devoted to assessing the project's impact or success. Typically a summative evaluation takes place after the project cycle has been completed and when it is possible that the impact of the project has been realized. It answers these basic questions as follows:

- Was the project successful? What were its strengths and weaknesses?
- Did the participants benefit from the project? If so, how and in what ways?
- What project components were most effective?
- Were the results worth the costs?
- Can the project be replicated in other locations?

Specific impacts that may be looked for, for example in a typical social development programme include health, nutrition, reproduction, child schooling, income, employment, etc. In addition, practitioners may want to know if the programme had any impact on poverty, women, empowerment, and domestic violence. Social science evaluators are challenged to construct impact assessments capable of measuring whether or not such a programme contributed to any gains in individual or family welfare. Nevertheless, a well-conducted summative evaluation helps decision makers—programme managers, donors, agencies—determine if the project is worth continuing. An honest evaluation recognizes unanticipated outcomes, both positive and negative, that come to light as a result of a programme.

3. *Other Types of Evaluation*

(i) Concurrent Evaluation

This refers to the evaluation of an on-going programme simultaneously with implementation. This is more or less similar to formative evaluation. Mostly used for monitoring the programme to assess interim goals/objectives, activities, flow of resources etc, to take corrective actions.

(ii) Utilization-Focused Evaluation

It is alternative goal-free evaluation. Evaluators routinely experience a lot of problems in attempting to state objective and goals acceptable to all the stakeholders involved in the programme. Also, the goal analysis, is often a complex, chaotic, controversial, and confusing effort involving time and energy. Some evaluators opine that goal-free evaluation is more suitable under circumstances like turbulent and more unstable environment of the programme; enables gathering of data on a broad array of actual effects and evaluating the importance of these effects in meeting the demonstrated needs. It does not discuss about goals with the stakeholders but discusses only about the programmes actual outcomes and measurable effects and these are judged on the extent to which they meet the demonstrated needs of the participants.

(iii) Formal and Informal Evaluation

These terminologies were used by Flutchy and Denoughy in the beginning of the Extension evaluation literature. These refer to the two terminal ends of the degrees of Evaluation Continuum described under the levels of Evaluation.

Levels of Evaluation

There are often several misunderstandings about the levels, degrees, and stages of evaluation. Basically the levels of evaluation refer to the evaluation at Individual, Household, Enterprise, Organizations/ Institutions, and Community/Society level. Sometimes these are interpreted as units of analysis. In some other cases, Simple, Middle Range, and Complex evaluation categories are also interpreted as levels of evaluation. Hence there is no consistency in the use of terminology in the literature about the levels of evaluation.

Degrees of Evaluation

Degrees of evaluation can be best explained on the continuum as seen below. It indicates evaluation categories based on the degree of formality in evaluation intensity or rigourousness of scientific method. This classification is often mistaken for levels of evaluation.

Casual everyday evaluation	Self-checking evaluation	Do-it-yourself evaluation	Extension studies	Scientific research

——————CONTINNUM——————

Stages of Evaluation

1. Pre-Evaluation Stage

Evaluation carried before the start of the project is called pre-evaluation of the project. This usually carried out with the intention of establishing a benchmark on the status of the unit intended for the project implementation.

2. Implementation Evaluation Stage

Evaluation carried out during the implementation of the project/ programme. Sometimes it is also called Concurrent Evaluation and is usually to generate information for monitoring the project activities and also obtain interim feedback on several issues that help in fine tuning the project and its implementation. This is similar to Formative evaluation described in this module.

3. Post Evaluation Stage

It is usually carried out after the project is completed and is similar to the summative evaluation to find out the outcomes of the project.

Types of Impact Assessment

When an impact assessment is planned, the type of assessment that should be used depends on the needs of the stakeholders. Determining these needs will define the type of tools and impact assessment that should be performed. Given below are the two most common types of impact assessments.

1. Donor-Led Impact Assessment

As mentioned earlier, donors require some evidence that their money is being used to effectively further their stated goals. A donor-led impact assessment examines the impact of a project or a programme from the perspective of the lender. Results of a donor-led impact assessment are often shared with the donor's funders, which are usually government agencies or foundations. Future funding decisions are often made based on this assessment.

2. Practitioner-Led Impact Assessment

While donor-led assessments have been the most commonly conducted assessments, there has been a shift towards practitioner-led assessments, which have a different focus. These assessments focus more on "how the impact assessment process can fit into existing work patterns, build on existing knowledge and experience, and produce results that can be easily used by the management.

3 Proving vs. Improving

According to Hulme (1999), donor-led impact assessment methods can be thought of as needing to "prove impact," while practitioner-led impact assessment is meant to "improve practice" of an organization.

Important Considerations for Impact Assessment

Effective evaluation depends on the ability of evaluators *to establish link between the changes identified during and after the course of the project, which are specifically attributable to the intervention.* It is not possible to conduct an evaluation without having any measures or indicators. Therefore, three standards for impact assessments have been established. They are *credibility*, *usefulness* and *cost-effectiveness.* This framework is designed to be flexible enough to take into account different types of programmes and different contexts. It also recognizes that there are necessary tradeoffs with these standards when conducting an evaluation.

1. Credibility

Credibility implies the trustworthiness of the entire evaluation process. It begins with clearly stated objectives that indicate the type of impacts that will be examined, the intended use of the findings, and the focused audience. The impact assessment should formulate a set of key hypotheses and seek to test those using quantitative and qualitative studies. The evaluation should establish and test a plausible relationship between the programme and changes as a result of participating in the programme. Credibility can be improved by using data gathering instruments that are well designed and clearly documented.

2. Usefulness

In order to be useful, an impact assessment must be designed to address the key questions and concerns of the intended audience. The usefulness of the assessment is enhanced when those who are expected to use the findings are involved in the planning, design, and analysis stage. Involvement by the main users of the data helps to ensure that their concerns are reflected in the impact assessment process. A key element of usefulness is the timeliness of the data. Impact assessment data also can be used to define strategic objectives, design and deliver appropriate products, and suggest new products. Finally, it can be useful for developing strategies to improve portfolio performance by reducing turnover, expanding outreach, and improving portfolio quality.

3. Cost Effectiveness

Designing a worthwhile impact assessment that provides the donor or practitioner with valuable information is often challenging, particularly when working with a limited operating budget. According to Carolyn Barnes and Jennifer Sebstad (2000), an impact assessment can be more *cost effective* if there is a good "fit" between the objectives, methods, and resources available to those assessing impact. If possible, relying on the successes or failures of past impact assessments can help in creating greater efficiency for an implementing organization. By learning which methods worked well and which did not, evaluators can learn from past mistakes to achieve greater efficiencies in the evaluation. These past experiences, or other examples in the literature, can be especially helpful in identifying meaningful and valid impact hypotheses and variables, developing data-collection strategies for obtaining reliable and valid data, and selecting appropriate analytical techniques.

Evaluation Designs

Under the best circumstances a well-designed evaluation enables the evaluator to say with confidence that the programme impacts were due to the programme interventions and not some series of outside factors, which happened to coincide at the same time. The ability of the evaluator to do this is based on the internal and external validity of the evaluation. Internal validity is the accuracy in concluding that the outcome of an experiment is due to the intervention. External validity is the extent to which the result of an intervention can be generalized. Good evaluation design controls for external factors that comprise threats to validity.

Threats to internal validity include :

- *History*—uncontrolled outside influences on participants during an evaluation;
- *Maturation*—processes within the respondents that may change their responses over time;
- *Selection*—biased selection of participants;
- *Testing*—the effect of testing on the results in an evaluation;
- *Experimental mortality*—changes in a group because some have left the evaluation study.

Threats to external validity include :

- *Multiple intervention interference*—several projects or interventions occurring simultaneously
- *Hawthorne effect*—knowledge about the evaluation
- *Experimenter effect*—changes in the respondent due to the presence of an evaluator-sensitization due to the pre-test or any part of the evaluation

It is not possible to control for all outside factors when conducting an evaluation. However, it is possible to increase internal validity by randomly selecting participants, randomly assigning them to groups and using a control group. A control group does not receive the intervention so that the effect of the intervention can be determined on the test group, which has received the intervention. External validity can be improved by careful adherence to good experimental research practices. Evaluators can choose from several types of evaluation designs to minimize threats to internal and external validity. In choosing the right evaluation design, evaluators will typically try to find a balance between internal and external validity, and cost-effectiveness.

1. Pre-Experimental Designs

This type of evaluation design controls very few external factors. Pre-experimental designs usually focus on a single programme and attempt to discover whether or not it has achieved its objectives. As a result, it is difficult to draw strong conclusions about the impact resulting directly from the programme, because evaluators cannot be confident if changes were caused by the intervention, or by a host of external factors. Even though pre-experimental design evaluations lack the sophistication of more advanced evaluation research, they are inexpensive and easy to use. Often evaluators are faced with situations where there is no baseline data and no control group. In these cases, pre-experimental design affords the best possible evaluation under the circumstances.

(i) Before/After Comparison

This design compares the same subjects before and after the programme intervention. There is no random selection of participants to the test group and no control group in this design. The before/after comparison is represented visually below :

O_1	X	O_2

Where, X is the intervention (project) and O is each observation. This design is simple to use and inexpensive, but it does not control for threats to validity. Because there is no control group, it is difficult to determine if any changes to the clients were a result of the intervention, (i.e., training or seminar), or some other factor, such as maturation.

(ii) After Only-One Shot Design

This design is used when the evaluator has no baseline data on the participants. The evaluator simply measures and observes how programme participants' score after the programme has operated for a set period of time. This design is represented visually below :

X	O_1

Where, X is the intervention (project) and O is the observation. The major limitation of this design is that the lack of an initial observation leaves little ground to infer whether the participants changed as a result of the (project) intervention. There is no random selection of participants. Evaluators, however, can strengthen this design by developing simple control groups.

2. *Quasi-Experimental Designs*

Quasi-experimental design is one that looks like an experimental design, but lacks the key ingredient-randomization. These types of evaluation designs are an improvement over pre-experimental designs because they attempt to control for and minimize threats to validity. In almost all cases, quasi-experimental designs incorporate control groups and thus, provide some standard of comparison to determine if the intervention (i.e., project activities/or training) had the desired impact.

Some common characteristics of quasi-experimental designs include :

- *Matching* is used instead of randomization. Evaluators try to find a group in a neighbouring town or region that would have similar citizen demographics to the experimental group.
- *Time Series* analysis is often involved. These designs are organized over time (Longitudinal) to examine changes in the participants over time.

One of the intended purposes for doing quasi-experimental research is to capture events over an extended period of time, as well as a large number of different events to control for various threats to validity.

(i) Time Series Design

The time series design helps to overcome the weakness of the one-shot, before/after, pre-experimental design. In this design, the evaluator looks at the data over a longer period of time, both before and after the intervention. The idea with time series data is to understand whether the programme caused a break in a more established pattern of trends, such as income levels or education and literacy levels, etc.

The graphic view of a representation of a multiple time series design consisting of a matching control group is given below :

O_1	O_2	O_3	X	O_4	O_5	O_6
O_1	O_2	O_3		O_4	O_5	O_6

Observations are made to both the control and experiment group. Selection of participants in a random manner is critical to the validity of this design. If gains are made in the experiment group which does not show up in the control group, the evaluator needs to eliminate all other plausible explanations before determining if the intervention (project) caused the change.

(ii) Non-Equivalent Control Group Design

This design is typically used when it is not possible to randomly assign subjects to groups, for political, ethical or other reasons. For example, when a microfinance project is evaluated, it may not be possible to deny some clients a microfinance loan, while giving it to others just for the purpose of creating control group. With a non-equivalent control group, an attempt is made to find a control group, which although not randomly selected, has similar socio-economic characteristics. This design is represented visually as given below :

O_1	X	O_2
O_3		O_4

Where, X is the intervention (project) and O is the observation. Evaluators using this design collect pre-test and post-test data and examine the results. This allows the evaluator to compare the two groups. If no differences are found after the programme, the evaluator can be skeptical about the effects of the programme. However, if there are differences, it is important for the evaluator to determine if the differences are due to the programme or some other reason.

3. Experimental Design

True experimental design utilizes random selection and control groups to ensure internal validity. It is often touted as the most rigorous of all research designs because the results are more "valid" than other designs, i.e., the data from the experiment measures what it is intended to measure—for example, the effect of credit for poor families in a microfinance scheme. The key to success for experimental design is the random assignment of participants to both the control and experiment groups. While no two groups of people will ever be exactly the same, random selection of participants improves what can be inferred from the intervention. Randomization means that each member of the population being studied has the same probability of being selected. This ensures that the population does not contain individuals with a characteristic that could influence the experiment results. For example, individuals who choose to become part of microfinance groups may have some difficult-to-measure characteristic, such as personal risk levels or entrepreneurship. Therefore, to understand the impact of microfinance on personal incomes it would be incorrect to only sample participants within the microfinance groups. The result would overstate the actual impact of micro-credit on personal income as the participants may be individuals whose incomes would have risen over time anyway because of their entrepreneurship skills. While experimental design may be the most desirable form of evaluation from a scientific perspective, it is likely to be difficult to carry out in most real world context. This design of evaluation is often intrusive and time consuming and may not be possible in all social science evaluations.

(i) Pre-test/Post-test Control Group Experiment

Evaluators using this design select a random group of participants and assign them to an experimental and control group. Random selection is the process of drawing a sample of people for the experiment. In order to obtain a random selection of clients, for example, a microfinance project evaluation; an evaluator might pick every 4th borrower from an alphabetical listing to get a random sample of 100 clients. 50 of these clients could be randomly assigned in a similar manner to the experiment and control group. The design is depicted in the following graphic presentation :

O_1	X	O_2
O_3		O_4

Here, both the control group and the experiment group have been randomly chosen and assigned. Participants receive the same pre-test and post-test checks. The goal of the evaluator is to determine if there were any differences in test results that could be attributed to the specific intervention. Random selection helps to control for factors like selection, as well as maturation and history.

(ii) Randomized Two Group Post-Test Only Design

Since some groups may show gains due to repeated testing, an evaluator might want to conduct an experiment where no pre-test is given. In this case, participants are randomly selected to an experiment and control group. The graphic presentation is given below :

O_1	X	O_2
O_3		O_4

Here, the intervention is administered (project programme) and the results are tested to identify differences in outcomes for both groups. The advantage of this design is that it is cheaper than a pre-test/post-test design. The addition of a randomly selected comparison group helps prevent threats to validity including history, maturation, and testing effects.

An overview of the three major types of experimental designs is given below :

Overview of Evaluation Designs

Parameters	*Pre-Experimental Design*	*Quasi-Experimental Design*	*Experimental Design*
Presence of a control group?	In some cases, but usually not	often	Always
Random selection of subjects from the population?	No	No	Yes
Random assignments of subjects and groups?	No	No	Yes
Random assignment of treatments to groups?	No	No	Yes
Degree of control over external factors	*None*	*Some*	*Yes*

How to Conduct an Evaluation and Impact Assessment

We have now discussed the importance and scientific rationale behind evaluation and impact assessments, as well as the factors that should be considered when approaching an impact assessment along with some evaluation designs. Next we examine the steps involved in conducting an evaluation.

*1. **Qualitative*** vs. ***Quantitative Methods***

Cost effective evaluation and impact assessments employ methods that are appropriate to answer the key questions to the degree and precision needed. The choice of the method to use for evaluation and impact assessment also depends on the purpose of the assessment and its audience. The typical evaluation or impact assessment can be conducted using two basic divisions of methods: qualitative and quantitative.

Common quantitative methods include sample surveys and semi-structured interviews. Common qualitative methods include focus groups, case studies, individual interviews based on key open ended questions, participatory appraisals, and participatory learning activities. Qualitative approaches can complement quantitative approaches and *vice versa*. The combination of qualitative and quantitative approaches can enhance an impact assessment. Key features and strengths of each of these methods and guidance for choosing the right method, are as follows :

Evaluation and Impact Assessment Methods and When to Use Them

Method	*When Appropriate*	*When Not Appropriate*
Sample surveys	• Project affects a large number of clients and statistical comparisons between groups over time or between geographic locations. Client population is heterogeneous and it is important to isolate the influence of factors unrelated to the project.	• Project has a small number of people and its implementation is recent and likely to have little impact.
Semi-structured interviews or mix of closed and open ended questions	• Purpose is to explore or document what changes are due to the programme—Time limited for development and testing of a survey based largely on close-ended questions—Users require both quantitative data and information that helps to explain the data.	• Users want information from a large number of individuals—Users want statistical comparisons between groups or locations.
Focus groups	• Priority is to understand motivations and perceptions of clients—Need information to interpret quantitative data and to understand the causal impact.	• Need information based on representativeness and to generalize information from participants to a larger population—Want to understand the socioeconomic level of project participants.
Case studies	• Want to search out rival explanations for unexpected or negative impacts—Need to illuminate and put a human face on quantitative data.	• Need information about large number of clients and want to generalize the information gathered to a larger population.
Participatory self-learning	• Programme's objectives include empowering clients. Attention given to community impacts.	• Do not have access to highly skilled persons to facilitate discussion. Priority is standardized and statistically representative data for a large and diverse programme population.

Most likely the evaluation method will be a mix of the various methods discussed and that follow. Qualitative methods address questions related to how and why and are best at uncovering reasons behind unanticipated results. Results are generalizable to theoretical propositions, not to populations. Quantitative methods address questions related to what, who, where, how many, how much, and the incidence or prevalence of a phenomena. The results may be generalizable to a larger population, depending on the sampling technique used. A brief outline of the steps that should be followed in both qualitative and quantitative impact assessments has been depicted in the previous page.

2. Planning Stage

Before an evaluation or impact assessment, there is generally a planning stage that ensures that it is approached in the most logical way. Some general recommendations for the planning stage could be as follows :

- When selecting a programme to evaluate, be sure that it is relatively stable. This will increase the likelihood that the evaluator will accurately examine the impact of the programme.
- Clearly state the *goals and targets* of the evaluation or impact assessment. The intended audience (donors, government agency, project implementing agency, including clients and other stakeholders) should be made explicit as this will drive the types of questions that are asked.
- Identifying the *type of evaluation design* that will be employed is appropriate at this stage. The choice of methods should be based upon the objectives of the evaluation, the key research questions, the availability of resources (time, money and people), and the degree of precision required and the possible application of results.
- State the *key research questions*. Again, these questions will be developed based on what information is to be collected and who is/are the intended user/s. The type of research questions will influence many of the other decisions including

the cost of the assessment. It is also useful to think about the *sample size* needed to conduct an assessment.

- Researchers require a large enough sample size to make sure that survey results closely resemble those of the true population. It is also essential to consider the unit of analysis—the individual, household, or institution.
- *Anticipate the level of effort and time-frame* needed to perform an evaluation. Allow enough time for designing the assessment, staff training, data analysis, as well interpretation and presentation of the results to various stakeholders.

After planning is complete, the next stage is the research design and implementation stages of the evaluation or impact assessment. These sections are divided into two major branches: qualitative and quantitative. The differences between these two approaches are emphasized in the next section.

3. Designing Stage

(i) Oualitative Methods

Qualitative methods are useful when it is necessary to understand relationships between variables. These methods promote in-depth examination of topics and are particularly useful to :

- Facilitate a better understanding of impacts and processes (why and how changes occur);
- Give a client perspective of changes due to programme participation and their value of these;
- Illuminate intangible or difficult to measure changes such as gender relationships,
- Reveal unanticipated consequences of the programme;
- Help clarify hard to interpret findings from a survey; and
- Allow clients to examine explain and learn from their experience.

Qualitative methods generally involve interviews and other fieldwork that does not necessarily result in uniform data that can easily be analyzed. When designing qualitative research it is important to involve the person(s) requesting the evaluation, as well as the key programme managers and other stakeholders who will be using the information. This helps to make sure that the interview questions and other tools are addressing their concerns. In addition, encouraging staff participation in the design of the assessment improves employee morale and provides a sense of ownership of the process.

Some of the tools used for qualitative analysis are :

(a) Focus group discussions with guided discussion topics to elicit information;
(b) In-depth "case-study" research looking at a range of impact questions or focused on specific issues (e.g., savings, empowerment, social capital, asset building, business development, intra-household relationships, etc.);
(c) Specific tools (e.g., interviews) to understand reasons for client drop-out; and
(d) Participant observation.

Plan for Selecting Participants: Select participants who are appropriate for the questions that need to be answered. The ability to read and write may be a prerequisite for some qualitative methods, such as a written evaluation. The location of a focus group meeting or interview may influence whether or not someone can participate in evaluation.

Data Analysis Plan. Once the qualitative data collection is completed, it will be necessary to analyze the information. It is important to have a clear plan in place to facilitate the analysis as soon as possible. The analysis plan ought to include a framework for organizing the information from the specific records following a format that reflects the underlying purpose of the qualitative study. This serves to summarize information across a number of interviews, focus groups or participatory sessions.

(ii) Quantitative Methods

Quantitative methods focus on numbers rather than words. Depending on the sampling technique and sample size used, the findings should be representative of the population beyond the individuals involved in the study.

When the goal of the impact assessment is to show that a programme caused certain changes in the target population, quantitative method is an appropriate research design. Using quantitative methods enables an organization to understand and track the impact of its programmes across a broad spectrum of clients. Quantitative techniques help to measure actual impacts of project inputs, while qualitative methods are meant to tell/help understand how and why these impacts are occurring.

There are several key steps involved in designing a quantitative evaluation or impact assessment.

(a) *Form the Hypothesis* : The hypothesis is a statement made in order to draw out and test its logical or empirical consequences.

(b) *Select the Variables* : In order to measure the hypothesis, variables need to be accurately identified and measurable. Examples of variables include income, education levels, marital status and number of children. Even a variable as straightforward as income will have measurement issues.

(c) *Design the Questionnaire* : When designing the questionnaire, it is essential to keep the intended audience in mind. Long and complicated questionnaires often frustrate respondents. Some questionnaires use a mix of open- and close-ended. Include standardized instructions for all field workers to establish some sense of uniformity to the data collection and recording. It is best to refer to a survey book as designing a useful survey is complicated.

(d) *Determine the Sample :* The larger the sample, the more confident evaluators can be that the survey truly reflects the population. The sample size depends on two things : confidence level and confidence interval. The confidence interval is the plus- or minus figure usually reported in opinion polls. The confidence level is expressed as a percentage and represents the percentage of the population within the confidence interval.

4. Implementation Stage

(i) For Qualitative Methods

The next step in the evaluation process is the implementation stage. In order to ensure that the evaluation goes smoothly, make sure that all logistics, such as transportation and materials are organized. For example, a focus group experience can be improved when the participants feel comfortable at a neutral location. When considering where to hold focus group meetings, Barnes and Sebstad (2000) suggest the following :

- A setting that provides privacy for participants;
- A location where there are no distractions;
- A non-threatening environment;
- A location that is easily accessible for respondents; and
- A place where seating can be arranged to encourage involvement and interaction.

Conduct a training session for those who will serve as facilitators, interviewers and/or recorders prior to the focus group. Training typically includes a mock interview to acquaint the interviewer with the types of questions and answers that may arise.

The final step in the implementation process would be the initial analysis of the data, as well as a write-up of the notes taken during the session. This should be done as soon as possible, to ensure that the information is accurately recorded.

(ii) For Quantitative Methods

As with qualitative assessment methods, some initial steps should be taken in case of Quantitative methods too, prior to implementation; which help facilitate the process. These include:

(a) *Plan Logistics* : This helps ensure research credibility, the timeline, and a cost effective evaluation. Examples of things to think about, ahead of time include :

- Schedule and arrange transportation for the data collectors;

- Coordinate the amount of time needed in each field site;
- Locate a place to train the enumerators;
- Arrange a place where core team members and enumerators can meet daily; and
- Ensure that supplies (paper, pens, and clipboards) and photocopying facilities are available.

(b) Selection of *enumerators* should be based on their previous survey experience, knowledge of the local language, ability to understand and write in the language used in the assessment, their availability during the survey period and willingness to work outside normal business hours. Their objectivity is important for collecting credible data. The gender of the enumerators should be roughly proportionate to the respondents, but quality and experience should not be sacrificed for gender balance. However, it is also important to consider if the gender, age, or ethnicity of the enumerators will affect response rates. In countries with a history of inter-ethnic hostilities, this is a sensitive issue and it should be thoughtfully addressed.

(c) *Train enumerators* to understand the purpose and objectives of the survey, how to introduce the survey to respondents, and the meaning of each question. They also should learn best interview techniques and how to record the answers.

(d) *Pilot test the questionnaire,* with a small number of both clients and non-clients who are not part of the sample. Close monitoring and coaching of the enumerators by core team members during the pilot test helps to ensure that the questions are clearly understood by the data collectors and respondents, the survey is introduced properly, and responses are recorded properly. A pilot test may reveal problems with the questionnaire that require changes. After this, the questionnaire and the instruction manual can be fine-tuned and, finally, photocopied for field use.

(e) *Supervision of enumerators* is critical for valid and reliable data. An impact assessment core team member should meet with enumerators on a daily basis to review their work, discuss problems, and provide feedback on the previous day's questionnaires. This needs to be done irrespective of whether or not the enumerators are programme staff.

(f) *Data entry* should begin as early as possible, ideally after the pilot test of the questionnaire. If a computer and electricity are available in the field, data entry can start after the first day of survey work and problems with the questionnaire and/or enumerators can be detected and rectified quickly. If data are being entered on more than one computer, it is important to establish that the files can be merged early in the process. Involvement of core impact assessment team members in data entry is an effective way for them to check the validity of the data and facilitate coding. It is important to document problems encountered during the implementation stage and how they were dealt with. This makes the process more transparent and provides a stronger basis for establishing the credibility of the study.

5. Data Analysis

Good analysis requires a dataset that is manageable in terms of quantity. It further requires sufficient resources (time, money and people) that have been planned for, ahead of time.

The *analysis of quantitative survey data* should focus on the key research questions and hypotheses and follow the initial data analysis plan. The first rounds of analysis should explore descriptive data, such as averages, and frequency counts and distributions for all variables by key analytic categories. The next round should involve simple statistical tests on data showing differences in the characteristics of the sample and in the impact variables. Subsequent analysis should look at clusters or disaggregations of interest (such as location, gender, and client poverty levels) to determine the extent to which impacts vary by group. If the quantitative survey data do not fit with the hypotheses, the analysis should proceed to explore what the data do reveal.

The *qualitative analysis* should be guided by the initial plan, but additional information and analysis that would illuminate the underlying purpose of the qualitative component should be included. After summarizing the findings, the information should be organized and meaningfully reduced by selecting and focusing on :

(a) Patterns and common themes on specific items;
(b) Deviations from the patterns and factors that might explain these; and
(c) Interesting stories that help illuminate the broader study questions.

Analysis of the quantitative data is likely to reveal issues to explore further through qualitative methods. Similarly, the qualitative findings also may suggest certain questions or variables to probe further in the quantitative analysis.

Triangulating the quantitative findings deepens understanding of impacts and helps to establish the reliability of the findings. (Triangulation is a term borrowed from the study of experimental methods and refers to any attempt to investigate a phenomenon using more than one method. It can be thought as a way to crosscheck the research. By combining research methods, threats to the internal validity of the data are reduced.)

6. Presenting and Interpreting the Findings

Presenting and interpreting quantitative and qualitative impact assessment findings should centre on key research questions, evaluation hypotheses, and questions that address rival hypotheses. It is important to define key terms and concepts and avoid the use of overly technical terms. Information on key features of the programme and context should be woven in, to illuminate and interpret the findings.

In presenting and interpreting *quantitative findings* :

- Provide a profile of the sample;
- Organize the other data to address the hypotheses and key research questions;
- Simplify the presentation by using a select number of tables, graphs, charts, etc.;
- Present the data clearly and accurately, and use averages complemented with distributional data to provide a fuller picture;
- Make it easy to see differences between clients and non-clients.

In presenting and interpreting *findings from qualitative components,* the purpose of the impact assessment should guide the approach. In general, qualitative findings should be integrated as appropriate with the quantitative data to illuminate and explain quantitative findings, address questions that complement the quantitative findings, and provide a human face and tell a story. In some cases discussion of qualitative findings may dominate, with the quantitative data used to generalize these findings to a larger population. In interpreting the findings, never draw conclusions without data to support them. Always support conclusions with data.

At a minimum, the impact assessment report should include :

(a) A clear statement of the objective of the impact assessment;
(b) A description of the programme including its objectives, products, methods, dropout rates and financial data;
(c) A description of the methodology used for qualitative and quantitative components, especially criteria and process for selecting participants;
(d) Findings, interpretation of results, conclusions, and when appropriate, recommendations for programme improvement;
(e) Difficulties encountered, how these were addressed, and lessons learned; and
(f) Appendices with more detailed information.

7. Dissemination

To be useful for decision-making, evaluations and impact assessments must be disseminated in a timely manner to the intended audience. Dissemination strategies may include bullet point summaries, power-point presentations, and strategic meetings. The time lag between data collection and presentation should be reduced to a minimum. While some assessors recommend a period not exceeding nine months, others argue for the advantages of immediate feedback. In fact, having programme staff directly involved in all of the stages is one way to ensure immediate feedback. In addition, an impact assessment linked to a larger programme evaluation is likely to reach a wider audience.

Additional Issues

For rapidly growing development organizations, staff resources will have to keep pace with growth and expansion pressures. Rapid growth requires hiring of more impact monitoring officers. As it moves forward, managers will have to continue fine-tuning the balance between the organization's evaluation and impact assessment needs and its staff resources. Similarly, organizations will have to seek the optimal balance between conducting in-house evaluations and monitoring or outsourcing these functions to consultants. Bringing everything in-house is an attractive, short-term way to lower costs. Also, standards that help an organization measure its performance are crucial. Financial audits can compare internal financial practices against established external standards. Developmental organizations can benefit from a set of common standards for "good practice" that would help assess impact performance just as they assess financial performance. Finally, as the organization presents its findings to clients it should ask for feedback through a combination of reports and focus group discussions. Maintaining a focus on clients' participation and feedback is critical if the findings are to be useful to them as well as to others.

REFERENCES

Baker, J.L. 1997. *Evaluating the Poverty Impact of Projects : A Hand Book for the Practitioners*. LCSDR/PRMPO. The World Bank. Dec. 15, 1997.

Barnes, C. and Sabstad, J. 2000. *Guidelines for Microfinance Impact Assessments*. A Discussion Paper for the CGAP3 Virtual Meeting, AIMS Management Systems International 600 Water Street, SW Washington, D.C.

Bassel, H. 1999. *Indicators for Sustainable Development: Theory, Method, Applications*. A Report of the Balaton Group, International Institute for Sustainable Development. 1999.

Darice Byrn et al; 1959. *Evaluation in Extension,* USDA. Publication.

Germanov, D; Irons, J.M., and Carver, J. 2002. *Evaluation and Impact Assessment. Document No. 56*. A US Department of State/NISCUP funded Partnership among the University of Washington, Evans School of Public Affairs, The Siberian Academy of Public Affairs and Irkutsu State University, The Russian MFP.

Hulme, D. 1997. *Impact Assessment Methodologies for Microfinance* : A Review Paper in conjunction with the AIMs project for the Virtual Meeting of the CGAP working Group, May 1997. Institute for Development Policy and Management. Washington, DC.

Lucas, H. 2002. *Good Practice in Development of PRSP Indicators and Monitoring Systems*. Working Paper 72. Results of ODI. Research Presented in preliminary form for Discussion and Critical Comment. ODI 111 Westminster Bridge Road. London, SE7JD UK. July 2002.

Patton, MQ. 1997 (Ed). "Beyond the Goals Clarification Game: Focusing on Outcomes". *Utilization Focused Evaluation: The New Text*. Sage Publications. International Educational and Professional Publisher, New Delhi. Ch. 7-8.

Sabstad, J. 1998. *Low-Cost Impact Assessment Methodologies for Microfinance Programmes*: A Discussion Paper for the Second Virtual Meeting of the CGAP Working Group on Impact Assessment Methodologies, Management Systems International, Washington, DC., 1998.

2

Indicators for Impact Monitoring and Assessment : Identification, Selection and Application

H. N. Byra Reddy

Introduction

Indicators are quantitative or qualitative variables that can be measured or described and, when observed periodically, demonstrate trends and help to communicate complex phenomena. An indicator is not the same as the phenomenon of interest, but only an indicator of that phenomenon. Understanding their role in the project *context* is highly complex. Output (performance) indicators help to monitor and evaluate a project's efficiency, by assessing whether planned activities or expected results were achieved within a given time and budget. On the other hand, Impact indicators are used to monitor and assess a project's effectiveness. They describe whether the outputs of a project had further implications, viz. *intended or unintended, positive or negative,* on the context and its population. Whether an indicator is considered a performance or an impact indicator depends on the formulation of the project goal, purpose and results. Rather than a clear-cut distinction, there is a gradual transition between these two types.

No single indicator can describe either the performance or the impact of a project sufficiently. The challenge, therefore, is to select *a set of impact indicators* that covers all important aspects of the context and that is manageable given the means and capacity of a project. Once a set of goals/objectives of the project have been agreed upon through a participatory analysis processes in the Project cycle management, the next step is to identify indicators—also in a participatory way—to measure progress toward those goals as a result of an intervention or a development project. The impact monitoring and assessment depend critically on the choice of appropriate indicators. Preferably, they should be derived from the identification and descriptions of relevant variables being given by the clients, with appropriate indicators of them being based on discussion with all the stakeholders. This chapter discusses on how to understand these aspects and how the indicators identified are applied in a practical situation for Impact Monitoring and Assessment (IMA).

Meaning of Terms Used in Impact Monitoring and Assessment (IMA)

Not all development organizations and references use terms related to project cycle management and impact monitoring in the same manner. Therefore, terminology used here is briefly described to clarify how terms are interpreted in the present module.

Goal (Overall Goal) : In a wider sense, the overall goal is the ultimate change desired in a *context*, e.g., poverty alleviation, sustainable resource management, empowerment of the local population, etc. The goal cannot be reached by a project alone, but a project should make a relevant contribution to the goal.

Expected Result and Output : The term "expected result" corresponds with the term "output", which describes a short- to mid-term result that is actually achieved as part of the responsibility of a project. Achieving outputs relates to the efficiency (functioning, performance) of a project.

Impact : "Impact" comprises the mid- to long-term implications a project has for the *context* and its population, be they *intended* (planned) or *unintended*. Even the presence of development workers or the mere existence of a project can have implications. Expectations are created, stakeholders may change their behaviour, etc., without a project having any input or conducting any activity. But as soon as a project is planned, the purpose and goal reflect intended impacts.

Therefore, "impact" is often related to the effectiveness of a project, i.e., its success in contributing to its goal.

Impact Chain: The term "impact" covers a wide range of implications, which can be seen as an impact chain of overlapping links. The utilization of project outputs already implies the idea of a broad impact (e.g., adaptation of a new crop production system with greater area coverage). As a consequence of utilization, initial effects (outcomes, direct impacts) can be observed (e.g., crop yield increases, soil erosion decreases, etc.). These effects may imply both benefits and drawbacks (e.g., increased crop yield must be marketable to increase household income). This can stimulate a learning process, people's attitudes and perceptions can change, and further (indirect) impacts may be triggered (e.g., local people gain self-confidence and further explore their potential). In the end, at least some of the impacts should relate to the overall goals of development cooperation (e.g., empowerment of local people, poverty alleviation, etc.).

Impact Monitoring and Assessment : "Impact monitoring" can refer to different instruments, such as environmental/social impact assessment (prediction) and impact studies (retrospective impact evaluation). "Impact Monitoring and Assessment" (IMA) is considered part of project's process of self-evaluation, an instrument of reflection and learning to better adapt project activities to a changing *context*. IMA comprises two aspects: observation (monitoring) and interpretation (assessment) of the changing context and the project's implications. Only a combination of both aspects provides a useful instrument for quality control in project cycle management. Monitoring should be done "objectively" to establish an information base. Assessment involves the "subjective" judgment of different stakeholders in accordance with their individual perceptions.

Attribution Gap : Outputs can be related to the expected results relatively clearly. But beyond that, the impact chain (utilization, effect, benefit/drawback, impact) needs time to develop, time during which the number of actors and their interactions increases. This makes it more and more difficult to attribute a change to a single factor or project. This is called the "attribution gap". Even with costly investigations, a project can only narrow, but not close this gap. Realistically, a project can only establish and show plausible relations between its actions and changes in the context.

Indicator: Indicators are quantitative or qualitative variables that can be measured or described and, when observed periodically, demonstrate trends; they help to communicate complex phenomena. A single indicator can describe neither the performance nor the impact of a project sufficiently. The challenge, therefore, is to select *a set of impact indicators* that covers all important aspects of the context and that is manageable given the means and capacity of a project.

For example: An agricultural project that helps develop improved crop production systems may use the measure "60 per cent of the farmers have increased their maize production by 20 per cent within 3 years" as a performance indicator to show its efficiency. But the same indicator also addresses some links in the impact chain, such as "utilization" of the outputs (broad impact, area coverage), and "effect" (production increase).

Classification of Indicators

1. Scientific Indicators

Scientific indicators tend to be measurable in quantitative terms; they are global within a given discipline and are meant to be comparable across space and time.

2. GrassRoots (Indigenous/Local) Indicators

GrassRoots (indigenous/local) indicators are signals used by local people (individuals, groups, communities) based on their own observations, perceptions and local knowledge, applied within specific cultural, ecological and spiritual contexts; they tend to be more descriptive.

Another classification of indicators says that, they can be broadly classified into two categories, namely, *final and intermediate*.

3. Final Indicator

When an indicator measures the effect of an intervention on individuals' say ' well-being', we call it a "final" indicator. For example, literacy may be considered one of the dimensions of 'well-being', so an indicator measuring it—say, the proportion of people of a certain age who can read a simple text and write their name—would be a final indicator. Sometimes final indicators are divided into *"outcome"* and *"impact"* indicators.

Impact indicators measure key dimensions of 'well-being' such as freedom from hunger, literacy, good health, empowerment, and security.

Outcome indicators capture access to, use of, and satisfaction with public services, such as use of health clinics and satisfaction with the services received; access to credit; representation in political institutions and so on. These are not dimensions of 'well-being' in themselves, but are closely related. They may be contextual. Thus, both the *impact* and *outcome* indicators should constitute the final indicators of impact assessment and monitoring impact.

4. Intermediate Indicator

When an indicator measures a factor that determines an outcome or contributes to the process of achieving an outcome, we call it an *"input"* or *"output"* indicator, depending on the stage of the process—in other words, an "intermediate" indicator.

Outputs differ from outcomes because the outputs are fully under the control of the project implementing agency that provides them. For example the number of schools built is an output, because it is directly under the control of education or other public authorities, while the number of children passing out of the schools is an outcome, because it depends on the behaviour of children and their families, which may be a resultant of the inputs and outputs in the project.

Although the main objective of the monitoring system is to track progress in outcomes and impacts, both final (outcome and impact) and intermediate indicators (input and output) should be tracked. Monitoring final indicators helps to judge progress toward the goals set. But final indicators (outcome and impact) are the result of several factors, many of which are outside the control of policy-makers and programme administrators. Intermediate indicators, on the other hand, generally change as a result of actions by the project and other agents. Moreover, final indicators generally change slowly over time, while intermediate indicators change more rapidly, giving an indication if not on what is happening with well-being, at least on what is happening with some of its determinants. This can make it possible to take corrective action while a programme is being implemented. Finally, information on intermediate indicators is often easier to collect.

Final and intermediate indicators should be complemented with other selected indicators to measure overall impact assessment and account for the *context* in which the project is being implemented, for example, rainfall or external demand for a country's goods, etc.

Features of Good Indicators

- Direct and unambiguous measure of progress/change.
- Measure factors that reflect the objectives.
- Varies across areas, groups, over time, and is sensitive to changes in policies and programmes.
- Not easily blown off course by unrelated developments and cannot be easily manipulated to show achievement where none exists.
- Can be tracked and is not too costly to track.

Identification and Selection of Indicators for Impact Monitoring and Assessment

Once a set of goals/objectives of the project have been agreed upon through a participatory analysis process, the next step is to identify indicators—also in a participatory way—to measure progress toward those goals as a result of an intervention or a development project. The impact monitoring and assessment depend critically on the choice of appropriate indicators. Preferably, they should be derived from the identification and descriptions of relevant variables being given by the clients, with appropriate indicators of them being based on discussion of all the stakeholders.

1. Facts to be borne in Mind while selecting Indicators of Impact Assessment

- Indicators should comprise comprehensive information about the programme outcomes.

- These indicators must represent all important concerns of all the stakeholders in the programme. An *ad hoc* collection of indicators that just seem relevant is not adequate. A more systematic approach must look at the interaction of the programme components with the environment.
- The number of indicators should be as small as possible, but not smaller than necessary. That is, the indicator set must be comprehensive and compact, covering all relevant aspects.
- Indicators must be clearly defined, reproducible, unambiguous, understandable and practical. They must reflect the interests and views of different stakeholders.

2. Procedure for Defining Indicator Sets

This deals with the practical procedure of finding appropriate indicators. The procedure has several distinct aspects and is as follows :

- *Conceptual Understanding* of the total component of the programme and the community/social system in which it is carried out (to be carried out). We cannot hope to find indicators representing the viability of the programme impact on the community/social system and its constituents, unless we have at least a crude, but essentially realistic, understanding of the total programme and its expected impact on the system and know what to look for. This requires a conceptual understanding—at least—a good perspective of the whole issue.
- *Identifying representative indicators* from a vast number of potential aspects of programme impact on the client community constituents. This means concentrating on essential variables of those programme components which plausibly have their effect on the totality of the community/social system.
- *Indicators should Quantify Basic Orientor Satisfaction :* We must arrive at statements about whether the effect of the programme on certain aspects of the system is threatened and, if so, how seriously. This requires translating indicator information into information about orientor satisfaction.
- *Participative Process :* Involve all stakeholders in this process of identification, definition and selection of indicators. Client involvement should be given a prime

importance. If the clients consist of several categories, care should be taken to ensure sufficient representation in the group constituted for participatory impact assessment and monitoring. It is advisable to consult more number of client groups (focus groups for each representative area), while defining, identifying and selecting indicators for Impact Monitoring.

These procedural steps require a large number of choices that necessarily reflect the knowledge and values of those who make them. It is essential to bring in a wide spectrum of knowledge, experience, mental models, and social and environmental concerns to ensure that a comprehensive indicator set is found for a given application.

3. Reducing the Number of Indicators to a Manageable Set

It will be obvious that a detailed analysis following this scheme will usually produce a large number of components of plausible impact, a long viability impact chains and many potential indicators. Furthermore, there will generally be several, appropriate indicators for answering each of the assessment questions or particular aspects of it. It is, therefore, essential to condense the system of impact analysis and the indicator set as much as permissible without losing essential information. There are several possibilities to do this. They are :

- *Aggregation :* Use the highest level of aggregation possible, because, when they are applied in the final impact assessment they are likely to be disaggregated into smaller components—according to the requirement of scheme of impact assessment.
- *Condensation :* Locate an appropriate indicator representing the ultimate effect of a particular programme activity, without bothering with indicators for intermediate effects.
- *Weakest-link Approach :* Identify the weakest links in the programme and define appropriate indicators. Do not bother with other components that may be vital but not related to direct programme effects.
- *Basket Average :* If several indicators represent somewhat different aspects of an orientor question, define an index that provides an average reading of the situation.

- *Representative Indicator:* Use indicator/variable that provides reliable information, characteristic of a whole complex situation. For example, instead of using a whole range of entrepreneurial behavioural measures, one can use a measure of achievement motivation (*n-ach*).

4. Participative Process for Identifying and Selection of Indicators in General

(i) Role of Scientific Method

Science can help significantly in assuring that the processes of indicator search, selection and aggregation are as objective and circumspect as possible. Science provides extensive knowledge and complex models in most fields. Even if this information does not and never will represent ultimate truth, it can be used to create an inventory and structure available knowledge. In particular, it is important to avoid an *ad hoc* collection of indicators. The choice should be based on a consistent theoretical framework supported by sufficient empirical evidence. In this way, systematic methods for indicator search and selection can be developed that can assure reliable results if carefully applied.

While, theory can provide a systematic framework for guiding the search for indicators and assessing for their validity and reliability, it cannot determine the final choice of indicators. This task remains to be completed by the investigators and it requires their subjective choice. The results will obviously be influenced by background, knowledge and experience of the investigators. It is, therefore, extremely important that people of different social and scientific backgrounds and political persuasion independently derive indicator sets by defining participating systems and use the scheme of orientor impact questions. The best we can do is to accept the unavoidable subjectivity and to make these processes as systematic, scientific and encompassing as possible, i.e., comprehensive, complete and reproducible. This requires transparency and reproducibility of the process, a compact and systematic approach, and comparability of the results.

(ii) Role of Experts and the Need for a Participative Process

Letting only a group of experts to a select of indicators in an area of development is, however, obviously the wrong method. Experts are likely to focus on issues and items of their professional expertise while neglecting others that may have a significant effect in the real system. A search for indicators can only be as complete and comprehensive as the imagination, knowledge and experience of the investigators allow. But the best knowledge of systems and problems, including their long-term perspective, can usually be found with those who have to cope with them daily: citizens, unemployed persons, small business, managers and administrators, farmers, doctors, social workers, police and educators.

Hence, this pool of intimate system and problem knowledge must be systematically included in the process of indicator search and their selection. In addition to this effort to cover the full spectrum of knowledge, a similar attempt must be made to represent the full spectrum of values. While available knowledge constrains the search and selection of indicators for values. It is, therefore, necessary to include all relevant views and value perspectives of a community in a participatory search and selection process.

(iii) A Participatory Process for Selection of Indicators for Impact Monitoring and Assessment

A participatory process for indicator selection is not a new idea. More and more development projects are using it. It is a necessity borne of the need to define an indicator set that can provide a full and complete picture of a problem situation or the viability of a given system or impact of an intervention. The search for a set of indicators of sustainable development can bring together citizens, administrators, business people and experts in a participatory process that strengthens sustainability-oriented planning and decision-making. However in case of an objective assessment of impact due to an intervention (project activities carried out) we can limit it to the stakeholders in a project—all those involved and affected by the intervention. The following steps provide, a brief description of the participatory process involved in identification and selection of impact assessment indicators.

Step 1: Convene a Working Group of Representatives of All Stakeholders Groups Representing a Broad Range of Views and Experience

The working group should consist of at least 10-20 but not more than about 25 people. They should represent the full spectrum of clients covering all categories involved in the project, reflecting the knowledge and values of the whole community for which the indicator set is sought. They should be selected for their ability to work together in a spirit of cooperation, despite their different views and backgrounds. They should be committed to meet regularly and to follow through with the process, even though it may take one or two years.

Step 2 : Define a Statement of Purpose

In the beginning of its work, the group must agree on and write down a statement of purposes of impact assessment. This statement should clearly delineate the objective in a way that can later serve to refocus the effort, should there be tendencies to shift the topic.

Step 3 : Develop the Values and Visions of the Group

Unless the group is clear about the spectrum of values and visions within the group from the beginning, there will be endless ideological discussions about every individual indicator later. The group must try to identify and write down the common values and visions about the project outcomes that are supported by all and which will be used later to select the indicator set. It must also record differences in values and visions between group members. They will later serve as critical test criteria for the comprehensiveness and completeness of the indicator set.

Step 4 : Review Available Data

Indicator sets should make as much use of available data as possible, but the available data should not determine the indicator selection! Survey data and regularly collected statistics about the project areas can provide a large part of the data even for a complex set of indicators that cover the range of values and views already expressed and listed by the group about project impact. Care must be taken, however, that indicators are not merely selected to fit available data. Often some indicators that are absolutely crucial for project impact assessment have never been collected in conventional surveys. For this reason, the drafting of an initial tentative indicator set, based on the method explained should precede the review of available data. This identifies an ideal set of indicators irrespective of their actual or potential availability. Every effort should be made to retain on this list of indicators that are deemed important, even though data are currently not available for all of them.

Step 5 : Draft an Initial Indicator Set

Based on the ideal set and information about available data, an indicator set is drafted by the group. The systematic viability assessment procedure explained earlier should be used at this stage to identify a compact and yet complete set of indicators. There are several important reasons for avoiding *ad hoc* selection: it is an extremely inefficient process; its results are unpredictable and shaped by spur-of-the-moment insights; the indicator lists produced in this may typically have large gaps in some important areas and are overly dense in others; the selection is often shaped by a few vocal participants; and it is not reproducible. While limiting the number of indicators to a manageable set, the group must ensure that all important aspects of impacts are represented.

Step 6 : Involve Client's Community Participation in Critiquing and Improving the Indicator Set

Once the group feels that it has developed a comprehensive set of indicators that fully represents its spectrum of knowledge and values, it must submit it to the wider community for critical review and improvement. This would involve publication in the local press, publication of brochures explaining the selection, public discussion, and public hearings. The feedback from this community participation serves to revise and improve the indicator set.

Step 7 : Involve Experts in Technical Review of the Indicator Set

Even an extensive and participatory search and selection process is no guarantee for a good indicator set. Many of the indicators may involve technical issues that only experts in their respective fields can adequately address. A technical review of the indicator set by experts is essential.

This is also true for the necessary check for completeness of the indicator set, based on the viability assessment procedure. However, the role of the experts should be very clear to all: they are called in to ensure precision, completeness and measurability of the set. They are not permitted to revise the indicator set according to their own limited view of the world.

Step 8 : Research for Required Indicator Data

Indicator sets should make as much use of available data as possible. Survey data and regularly collected statistics can usually provide a large part of the data even for a complex set of indicators as already explained. However, it is highly unlikely that data for all indicators of the set will be readily available. Often it will be possible to use combinations of existing statistical data to quantify a particular indicator. In this phase of the effort, the sources for the required indicator data have to be identified, and measurement programmes for missing data have to be defined. Obviously, if it is impossible for organizational or financial reasons to obtain data for a particular indicator, a more accessible replacement will have to be found. But expediency should never be used as an excuse for not adopting and measuring an indicator for an aspect that is deemed essential.

Step 9 : Publish and Promote the Indicator Set

Once the final indicator set is adopted, and means of quantifying all the indicators have been found, every effort must be made to ensure that it will be used by all sectors of society for the assessment of current conditions and the guidance of policy. The indicator set and the sources of information for it must be published and made widely available. It should be actively promoted to ensure that it will be used as a reference for public discussion of issues of and related contexts. It is not necessary that the indicator set is formally adopted by political and administrative bodies. In fact, use of the set by Non-Governmental Organizations as an instrument for checking and controlling government and administration may be much more effective.

Step 10: Review and Update the Indicator Set in a Transparent, Formal Process

Impact implies constant change, and the indicator set itself will have to be adapted to changing conditions. It is, therefore, necessary to provide for periodic review and update of the indicator set, its database and the values and visions underlying its conception. It seems advisable to review and update the database every two years, to have a technical review every five years, for the values and visions, and to update the indicator set accordingly.

Final Selection of Impact Indicators

The impact chain can be of great help during the selection process. An existing indicator may already address one of these aspects and can thus serve as an impact indicator. Beyond that, additional impact indicators need to be found. Possible Impact chain resulting from new production and conservation technologies (output), and corresponding impact indicators is given below :

Links of the Impact Chain	*Impact Chain (Positive and Negative Implications)*	*Possible Impact Indicators*
Utilization of outputs	Farmers in the project area apply new production and conservation technologies (applicability) and adapt them to their specific situations (adaptability)	• % farmers adopting technologies without incentives
Effects	Crop production increases/ decreases, pests and diseases are minimized/increased, soil degradation decreases/increases	• Crop yield • Occurrence of pests and diseases • Soil erosion
Benefits/Drawbacks	Improved agricultural production is (not) marketable, household income increases (decreases), and women's economic position is strengthened (weakened)	• Household income • Women's labour income
Impact	Men and women decide jointly (men decide) how to reinvest household income; farmers experiment more (less) than before; soil fertility improves (decreases); more (fewer) boys and girls attend school	• Household decision-making • % of farmers experimenting with cropping practices • Soil fertility status

In addition to several procedures suggested earlier, we may supplement the following principles and examples, in identification and selection of IMA indicators. It is only to emphasize that selection of these indicators is as important as the entire process of IMA itself. The aim of IMA is to achieve a reasonable quality of information in order to find reliable connections between the project and changes in the context.

The following principles and the guiding questions can help in selection of impact indicators :

Principle	*Guiding Question*
Relevance	Is the indicator essential, i.e., does it really provide the information required for making relevant decisions?
Reliability	Is there a need for quantitative or qualitative indicators for a decision?
User-orientation and transparency	Is the indicator understandable and meaningful for the relevant stakeholders (land user, policy-makers, etc.)? Are there local indicators that can be used?
Feasibility	Does the project or the stakeholders have the means, skills and time to monitor the indicator?
Gender-orientation	Does the indicator bring to light gender-specific knowledge and issues?
Hierarchy/Area coverage	Do all indicators reveal changes at the same spatial/ decision-making level (field, household, community, catchment, district, etc.)?
Sensitivity	Is the indicator sensitive to short-, mid-, or long-term changes?
Sustainability-orientation	Do the selected indicators represent all dimensions of sustainability (social/institutional, economic and ecological)?

But not all indicators that are identified can be monitored. The project's means, time and resources on the one hand, and the stakeholders' interests in IMA on the other hand, will lead to a final selection of impact indicators. It should be kept in mind that these indicators are the basis for, but, not the only source of valuable information. Systematic monitoring can always be combined with gathering and documenting information from statistics, newspapers, discussions with partners, consultants, and informants, one's own observations and the like. There is no need to wait three years for the first results of the impact monitoring. For example, market prices of cereals and their fluctuations could also be determined by project staff while shopping for their families. Negative developments in the agricultural sector will come out during talks in a village or with colleagues. Such information can always be documented and serve as a background for an interpretation of changes at a later stage.

Indicator Sensitivity

Since sustainability implies a long-term perspective, each indicator should be checked to determine whether it is sensitive to short-, mid- or long-term changes. Indicators of short-term sensitivity (1-3 years) will be highly relevant for Impact Monitoring and Assessment as part of the project's self-evaluation process. They are helpful for immediate correction of project activities that are taking a negative direction and can also be monitored over a long period. Indicators that are not sensitive to short- and mid-term changes are more important for monitoring far-reaching or late impacts. They only help the project to adjust its activities after 5 years or more.

Preparing for Application of Indicators to Impact Assessment

In preparing for impact assessment, following details need to be considered :

- Ideally, all stakeholders agree on a common rating for all impact indicators. But it can also be interesting to carry out impact assessment separately for each stakeholder group, and each group's findings will be communicated to the others.
- It should be determined at what level the assessment will be made (household, community, etc.). For example, if there is a great heterogeneity of household categories (such as poor and wealthy households), changes in their context should be assessed individually or at least separately for each household category. If all households are judged together at the community level, the result may not reflect important changes in individual households. It would thus be meaningless!
- After a set of impact indicators has been selected, an initial observation (monitoring) that takes all of them into account

produces the baseline. In the first years to come, monitoring and assessment will only include those indicators that are sensitive to short-term changes. Indicators sensitive to mid- or long-term changes will gradually be added.

The Baseline Dilemma

Indicators not only represent components of a project context; they are also a means of communication between stakeholders. Thus, they must be selected jointly. On the one hand, it is recommendable to have a set of indicators fixed as early as possible, because it helps to establish a baseline, particularly for long-term observations. On the other hand, there are good reasons to take time with the selection. For example, the project context and the stakeholders cannot be well known and understood in the beginning. During the lifetime of a project the context and the views of the stakeholders change, and so may the indicators. Some of the initially selected indicators may become impractical to observe and need to be replaced. Furthermore, unexpected impacts may require additional indicators at a later stage. But sound indicator selection only at the end of the project is too late. As a compromise, several months should be dedicated to a participatory search for a set of impact indicators, for adapting the initial choice, and for incorporating "emerging" indicators. This is important because it documents the learning process of a project and its stakeholders. Indicators can always be added, but a basic number of indicators should be found, say after six to twelve months, to ensure long-term monitoring.

Conclusion

What are indicators, how can we classify these indicators based on the purpose for which they are used to measure and monitor different components of impact, the basis for identification, selection and how to apply them in IMA have been discussed. Demonstrating how the different kinds of indicators are applied in the actual process of Impact Monitoring and Assessment need to be considered separately. Also, the choice of appropriate methods of monitoring and assessment that are compatible with different kinds of indicators to yield useful and credible information to assess the impact are to be considered very judiciously and through stakeholders participation. Finally, one must understand that selection and formulation of impact indicators cannot be achieved in one short planning session. It is a process of optimization that may take several months.

References

Bassel, H. 1999. *Indicators for Sustainable Development : Theory, Method, Applications.* A Report to the Balaton Group, International Institute for Sustainable Development (IISD), 161 Portage Avenue East, 6th floor, Winnipeg, Manitoba, Canada. 1999.

Estrella, M., Blauert, J. *et. al. Etd. Learning from Change : Issues and Experiences in Participatory Monitoring and Evaluation.* Intermediate Technology Publications, IDRC, Canada.

Feder, G. and Slade, R.H. 1986. "Methodological Issues in the Evaluation and Extension Research." In Etd. Jones, G.E. 1986. *Investing in Rural Extension: Strategies and Goals.* Elsevier Applied Science Publishers, London.

Herweg, K. and Steiner, K. 2002. *Impact Monitoring and Evaluation: Instruments for Use in Rural Development Projects with a Focus on Sustainable Land Management. Vol. 1: Procedure.* Rural Development Department, World Bank, Washington D.C. 1998.

Patton, M. 1997 Etd. *Utilization Focused Evaluation : The New Century Text.* Sage Publications, Chapters 7-8. International Educational and Professional Publishers, New Delhi. 1997.

Prennushi, G., G. Rubio and K. Subbarao. 2001. *Monitoring and Evaluation A Draft for Comments.* World Bank Washington, D.C.

Sustad, J. and M. Cohen. 1998. *Toward Guidelines for Lower-Cost Impact Assessment Methodologies for Micro Enterprise Programmes.* A Discussion Paper, AIMS, Management Systems International, 600 Water Street SW., Washington, DC.

3

Measurement of Social Impact through Participatory Approaches

K. Narayana Gowda

The biggest challenge facing the social scientists across the countries is measurement of changes on social counts due to implementation of a development programme. The reason being, social impact is not easily observable and they are qualitative in nature when compared to economic outcomes/indicators. Further, the tools required to measure the social indicators are not readily available and even if they are available not suitable to capture the diverse situations.

What is Social Impact?

Social Impact is designed to assist communities in making decisions that promote long term sustainability on healthy community and social well-being.

It is defined as changes that occur in people's way of life (how they live, work, play and interact with one another on a day-to-day basis); their culture (shared beliefs, customs and values); and their community (its cohesion, stability, character, series and facilities).

It is also the process of analysing (predicting, evaluating and reflecting) and managing the intended and unintended consequences on the human environment of planned interventions (policies, programmes, plans and projects) and any social change processes involved by those interventions so as to bring about a more sustainable and equitable, bio-physical and human environment.

Why Social Impact Assessment ?

Social impact assessment is important for several reasons. It is used to alert the community, including residents and local officials, the extent and magnitude of the proposed development on the community's social well-being. The assessment can help communities avoid creating inequalities among community groups as well as encourage the positive impact associated with the development. It provides estimates of expected changes in demographics, housing, public services, aesthetic quality of community and diverse community values. It also provides a foundation for assessing the cumulative impact of development on community resources and also to prioritise the needs. It helps to modify the planned interventions to reduce the negative impact and enhance positive impact. Unfortunately, social impact assessment takes a backseat to other types of impact assessment such as fiscal and environmental aspects because the impact are often more difficult to quantify and measure. Social impact assessment should be an integral part of the development process.

Who should be Involved ?

Since the impact assessment is designed to estimate the effects of development programme on community's social development, the process should rely heavily on involving community members who are influenced by the development. Human beings are at the centre of concerns for sustainable development. Others who should be involved in the process include community leaders of different background, representatives of various local institutions and officials who are involved in the programme/project implementation. In other words, the stakeholders who are involved directly and/indirectly will constitute the respondents for social impact. This should be flexible keeping in view the local needs and type of development programme.

What are Social Impact Indicators ?

Social indicators are sign posts that help measure progress towards a desired outcome. Indicators are chosen because they measure the outcome of interest directly. It is difficult to give comprehensive social indicators for a development programme assessment; it varies with the type of programme and its duration, type of clientele, stage of development and so on.

Indicators have to be selected against the following criteria (first established in Social Report 2001)

- *Relevant to the Social Outcome of Interest* : The indicator should be the most accurate statistic for measuring both the level and extent of change in the social outcome of interest, and it should adequately reflect what it is intended to measure.
- *Based on Broad Support* : Ideally there should be wide support for the indicators chosen so they will not be changed regularly.
- *Grounded in Research* : There should be sound evidence on key influences and factors affecting outcomes.
- *Able to be Disaggregated* : It should be possible to break the data down by age, sex, socio-economic status, ethnicity, region and where possible, to the individual (or smallest group possible), so we can compare outcomes for different groups.
- *Consistent Over Time* : The usefulness of indicators is related directly to the ability to track trends over time, so indicators should be consistent over time.
- *Statistically Sound* : The measurement of indicators needs to be methodologically rigorous.
- *Timely* : Data needs to be collected and reported regularly and frequently to ensure that indicators are providing up-to-date information.
- *Allow International Comparisons* : Indicators need to be consistent with those used in international programmes so we can make comparisons.

Realizing the importance of social dimension of development, an International Committee recommended the following items to constitute the indicators of development :

1. Health including demographic features, food and nutrition.
2. Education including literacy and skills, conditions of work, employment situation, aggregate consumption, savings, transportation.

3. Housing including household facilities, clothing, recreation and entertainment, social security, human freedoms.

Development effects the structure and functioning of the society in many ways. Developed societies are characterised by low mortality and fertility rates with low population growth while underdeveloped societies suffer from high mortality, high fertility and high population growth.

In the context of agricultural development programmes, the social indicators vary with the type of programme subjected for impact assessment. Some of them are immediate benefit indicators and others are long term benefit indicators. Yet another way of looking at social indicators are direct and indirect benefits as well as desired/undesired consequences. Social like change in knowledge, skills, adoption, perceptions, needs and aspirations, social status, leadership development, decision-making pattern, activating and facilitating the performance of grass root level institutions, minimizing inequalities, promoting new local institutions, improving nutrition status and standard of living, etc., could be considered while studying impact of agriculture and rural development programmes.

How to conduct a Social Impact Assessment

The impact assessment needs to be done at two phases:

1. Defining the scope of impact assessment, and
2. Identifying and evaluating the impact.

1. Defining the Scope

Carefully defining the scope of impact assessment can save scarce resources such as time, money and materials. Defining the boundary of social impact assessment should be consistent with mandate of the project/programme.

2. Identifying and Evaluating the Impact

Listing of appropriate indicators, planning for execution in terms of location, sample size, detailed method of data generation, analysis through appropriate statistical tools and report preparation, are some of the important activities to be carried out during this phase.

Techniques of Impact Assessment

Traditional survey method of data generation was extensively used, Thereafter, structured interview method was employed for a long period for social impact assessment. Observation methods such as participant or non-participant methods were used particularly in anthropological studies. In the recent years, participatory methods are being employed for social impact measurement.

What is Participatory Approach for Impact Assessment ?

Involvement of all sections of people is the crux of the problem in any assessment process. The participatory approaches can overcome these deficiencies and, therefore, they are used as the appropriate alternatives for quality, reliable and accurate data generation. Participatory methods are necessary for rapid exploratory assessment. Participatory methods for impact assessment are most commonly associated with the spread of diagramming and visual techniques. By the end of 1980s, diagramming techniques of bringing together the insights from different disciplines had combined into a flexible methodology referred to as Rapid Rural Appraisal (RRA).

Initially, the term PRA used to describe the bringing together of PRA and activist research. These were flexible application based on a number of underlying principles.

- *Embracing complexity* and seeking to understand it rather than oversimplifying reality in accordance with predetermined categories and theories.
- *Recognition of multiple realities* to be taken into account in analysis or action.
- *Prioritising the realities of the poor and most disadvantaged* as equal partners in knowledge creation and problem analysis.
- *Grass roots empowerment* aiming not only to gather information about impact, but to make the assessment process itself a contribution to empowerment through linking grass roots learning and networking into policy making.

More recently the term Participatory Learning and Action (PLA) has become preferred because it more effectively incorporates the underlying human rights tradition through emphasising the importance of:

- Changing from *appraisal to learning* and hence moving away from the use of participatory methods as an extractive process by outsiders to a sustainable learning process involving different stakeholders as equal partners.
- *relating learning to action* incorporating programme and policy improvement as an integral part of the learning process.

Participatory Techniques

The participatory techniques could be role play, flow diagrams, venn diagrams, time trend analysis, mapping, voting, seasonal calendars, wealth ranking and other ranking techniques, triangulation with other methods, photography, video and grass root diaries, proverbs, local songs and any other creative.

Participatory methods are, therefore, a diverse and flexible set of techniques for visual representation and stakeholder involvement characterized by a set of ethical principles. There is no one set of techniques to be mechanically applied in all contexts for all participants. Generally, both visual techniques and participatory facilitation are combined in different ways. The emphasis is on innovation and creativity in adapting previous practice to new contexts and needs.

Advantages of Participatory Methods in measuring Impacts

- Rapidly identifying the range of potential impact.
- Participatory prioritisation of different impact goals.
- Identification of locally relevant indicators.
- Identifying relevant stakeholder categories for assessment, control groups and analysis.
- Involving different stakeholders in a participatory process, including the most vulnerable.
- Rapidly obtaining impact and other information for whole communities or groups.

- Exploring processes and hypotheses.
- Rapidly investigating underlying or side issues.
- Increases communication between donors, policy-makers, development practitioners and those affected by interventions.
- Makes information immediately accessible to different stakeholders.

Limitations

- Difficult to extract sensitive information.
- Full knowledge of local language on the part of moderator.
- Difficult to find timing, place convenient to all particularly when they are scattered.

Conclusion

Participatory methods can be used on their own for rapid exploratory assessment in order to increase their reliability and credibility. They need to be *triangulated* with other quantitative and qualitative methods as part of an integrated social impact assessment.

References

Burdge, R.J. 1995. *A Community Guide to Social Impact Assessment*. University of Illinois: Urbana.

Chambers, R. 1994. *Participatory Rural Appraisal: World Development*. 22(7): 953-969.

Parshad, R. 2005. Impact Assessment of Planned Interventions. Framework and Approach. Paper presented in National Seminar on Methodological Issues in Impact Assessment of Agricultural and Rural Development Programmes.

4

Participatory Poverty Assessment: Can PRA be an Effective Tool ?

P. Adhiguru and *C. Palanivel Murugan*

Introduction

Nearly one billion of world population is still under the clinches of poverty. This segment of the population is generally, poor, less educated, unemployed or under-employed and has less access to most of the essential resources of livelihood. In many cases, the poverty goes hand-in-hand with many forms of deprivation, including deprivation of material resources, human resources, social contacts, voice and power (Kozel and Parker, 2000). Poverty is characterized not only in terms of insufficient income or absence of employment opportunities but also as complex syndrome, which manifests itself in many different forms that need a holistic approach to understand its nature and change. Despite the number of initiatives taken on poverty alleviation, the poverty situation remains unchanged in most of the developing countries.

It has been realized that due emphasis has to be given for assessing poverty precisely so as to build effective anti-poverty strategies. In general, household income and per capita expenditure are the yardsticks used in measuring the standard of living or poverty. There are, however, strong criticisms that per capita income and expenditure do not capture dimensions as health, literacy and access to public resources like safe drinking water and clean environment (Hadi and Chowdhury, 2001). Further, the data collection methods that are commonly used to measure poverty miss critical aspects such as behaviour not revealed by the market behaviour, the impact of family characteristics on well-being and the welfare effect of public programmes (Ravallion, 2002). Thus, quantitative approaches, widely used in poverty impact monitoring, have their own limitations of having a partial view of this complex, diverse, and area-specific realities in which the poor lives. As a result, qualitative approaches are though gaining significance in poverty assessment, but are being viewed by many researchers as having certain limitations like generalization and comparisons across regions or countries.

Therefore, in poverty impact assessment, no single methodology is considered ideal in terms of comprehensiveness, accuracy, cost-effectiveness and time efficiency. It has also been suggested that for a comprehensive understanding of poverty in a given region or country, a combination of qualitative and quantitative techniques would probably be the best. This would combine the advantages of statistical quantification with the detailed insights of the qualitative participatory approach (Haan and Laier, 1997).

Because of the widespread anomalies in the poverty impact measurement methodologies, the World Bank introduced a new approach, generally known as 'Participatory Poverty Assessment' (PPA), which uses both quantitative and qualitative methods. The World Development Report 1990 and subsequent policy papers by the World Bank projected PPA as an important process for identifying problems of the poor and linking them to the broader policy issues. The Poverty Impact Monitoring (PIM) studies are increasingly concerned towards poor in terms of heterogeneity with respect to their livelihoods, well-beings, vulnerability, etc. It emphasizes the need of using participatory poverty methods in revealing the hidden poverty dimensions.

Based on the above background, the present study was undertaken with the broader objectives of exploring the scope of using various Participatory Rural Appraisal (PRA) techniques in Participatory Poverty Assessment (PPA) research studies.

Objectives

The specific objectives of the study were:

- To enumerate PRA methods used in Poverty Assessment studies and explore the scope of using them in various contexts of poverty,
- To illustrate specific cases for which PRA methods could be a complementary tool in measuring poverty,
- To conceptualize a Participatory Poverty Monitoring System, and
- To enlist strengths and weaknesses of PRA.

Methodology

The study was based on reviewing of the past research experiences in the field of Participatory Poverty Assessment. The opinions of different eminent researchers in the field of PRA were also collected through informal discussions and were used to elucidate the poverty conditions in which the PRA could be best used to measure different poverty levels. Besides this, the case study approach was employed to explain the uses of PRA in PPA studies.

PRA Methods and Poverty-related Indicators

Various PRA techniques are widely used in PPA studies to understand causes, effects and dynamics of poverty. New PRA tools are being framed and applied to meet out the demands of measuring poverty more precisely. Types of poverty indicators to be measured and the situations under which these tools are applied determine the appropriate PRA tool to be exercised. Some PRA methods are found to be complementary to other PRA tool or are used in conjunction with quantitative data collection instruments. This kind of parallel application of PRA techniques is mostly admissible in the context of triangulation, a concept to cross verify the facts under a situation. Thus, it adds value to the generalization of a problem situation with more authenticity. Some PRA methods along with poverty-related indicators are described in Table 4.1.

Scope of PRA in Participatory Poverty Assessment

Participatory appraisals are meant to deepen the understanding of poverty, particularly people's own perception of the situation, often

Table 4.1: Some PRA Methods and Poverty-related Indicators

PRA Technique	*Indicator*
Social map	Social structures, social stratification, social facilities, etc.
Resource map	Resource availability and level of access by households
Mobility map	Nature of movement of household members, and their constraints
Time line	Sequence of natural events, interventions, social events and the developments thereafter
Time trend	Time series on yield, price, population, area, etc.
Livelihood analysis	Sources of food, income, food insecurity, etc.
Matrix ranking	Categories of households with their preferences, needs, problems, etc.
Wealth ranking	Grouping of households based on indigenous wealth indicators
Venn diagram	Social relationships among institutions, key individuals
Daily routine diagram	Nature and time share of routine activities
Transact	Geo-physical resources
Impact diagram	Positive and negative influence on poverty levels
Flow diagram	Establishing cause and effect path of variables/factors

crucial for policy making. PRA has unique role in Participatory Poverty Assessment. It addresses relevant and important questions, such as :

- How people define poverty in a locality? What are their perceptions?
- What kind of criteria will be more appropriate to measure poverty?
- Is there any variation across individuals, households and communities? Are there any changes in perceptions over the period?
- Whether appropriate questions are put before people whom we are studying?

In this context, intense debate on participatory approach for poverty assessment has been taking place at various world forums. Moreover, the acceptance of combined use of both qualitative and quantitative methods for comprehensive poverty measurement has been appreciated in various research findings. On account of the inherent merits of PRA methods, the researchers on poverty feel that, these should be incorporated in poverty measurement surveys/programmes to estimate the poverty in its real sense. With this background, it is intended to conceptualize a very elaborate methodology, which use PRA individually as well as combined with quantitative methods for a comprehensive poverty measurement. The proposed poverty assessment process (Figure 4.1) has different stages like problem identification, determination of sample size, nature of sample and location of sample households, data collection, data analysis and final results. It also suggests some of the relevant PRA techniques, which could be used independently or in conjunction with quantitative tools.

PRA and Different Dimensions of Poverty

Various conditions under which PRA would be of greater use in Participatory Poverty Assessment are highlighted below.

Poverty often goes hand-in-hand with illiteracy, which limits the understanding and interpretation of logic. The illiterate poor are less aware of the causes of poverty or otherwise are not able to explain them appropriately. Further, they cannot directly explain some of the standard indicators, especially when it involves minor calculations. e.g., relative income loss or gain. In such cases, PRA is the most appropriate methods, because of its dramatic nature and this helps them in identifying and explaining the causes.

The indigenous variables that are more essential and can be used as proxy poverty indicators are very often not cared by the quantitative surveys. Thus, actual poverty levels as perceived by the poor may not be assessed. The PRA has unique role in identifying indigenous variables in the given contexts and locations. For instance, only landholdings alone cannot be considered as proxy indicators of poverty, but also, the characteristic of the holding (e.g., fertility status) may differ and may make differences in poverty levels. In such cases, use of relevant PRA methods like focus group interaction and key informant's opinions provide information on indigenous variables like land characteristics and their links to poverty levels.

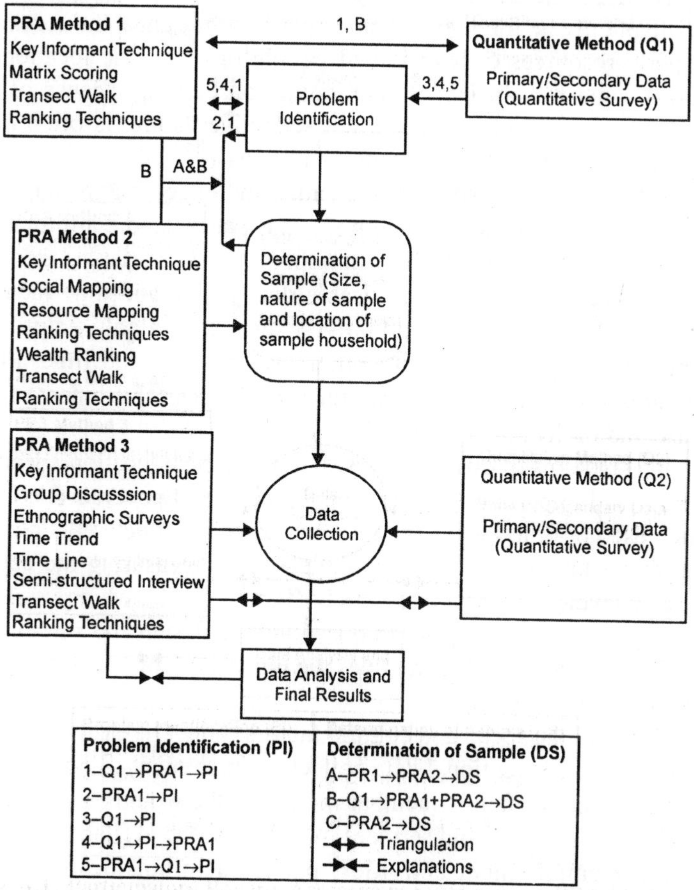

Fig. 4.1: Participatory Poverty Assessment System using PRA Methods

Poverty assessment in terms of assessing community effects on poverty status warrants locations of different community households. Under such cases, if one directly goes for quantitative surveys, it may not be time and cost effective. However, pre-survey mapping of community households using PRA methods, would be highly helpful in determining appropriate sample design and thus saves reasonable cost and time. For instance if three community households are there in a village for a total sample size of 90, the equal sample distribution of 30 can be surveyed for each community by using prior community household mapping.

Homogeneity of response with respect to a group or a community may explicitly be foreseen in some cases. For instance, the common property resources endowment of people of a specific location is almost same for all. Under such conditions, PRA, because of its comprehensive, collective, and cost-effective nature, is often best suited to handle the homogeneity conditions.

A better understanding of the cause and effect relationships, as predicted by the quantitative methods, may warrant more information on variables in question. Re-examining of these variables through quantitative surveys is not advisable due to cost and time factors. Use of PRA, as a post-quantitative exercise, is highly helpful in explaining the variables with relatively less cost and time. PRA provides new impetus in meaningful and logical establishment of the cause and effect relationships.

The patterns of different causes, most of them location-specific, were found important in explaining the degree of poverty. The PRA when used prior to the quantitative survey could effectively uncover the above factors and increase the scope of measuring the actual poverty levels. For instance, a study (Barret, 2001), which used the Participatory appraisal prior to questionnaire design and survey site selection, underscored such issues the importance of which they had not previously appreciated.

Some of the social and cultural factors cannot be easily captured by using the conventional data collection instruments. For instance, as the number of family member increases, the woman has to spare more time for household services. As a result, the availability of her time for employment activities, particularly outside the house decreases. Similarly, households with more children and or children often falling sick, limit the income of the woman members of the households. The healthcare services of these children rob the time available for outside employment activities and also need more expenditure. Consequently, there is demand for more income. This combined effect pushes the household, especially women, to transient poverty. Such hidden dynamics can be comprehensively assessed by the PRA methods. Besides making the poverty assessments appropriately, PRA, when adopted in action-oriented research, is helpful in encouraging the people's involvement in its implementation phase.

PRA and Different Dimensions of Poverty : Some Case Studies

Specific conditions under which PRA can play a role in poverty measurements are explained through the following three case studies.

Case I

Price Heterogeneity and Measurement of Inequality

The term 'poverty' may itself be misleading if it suggests that all low income households share a common set of characteristics. In fact, the poor constitute a 'heterogeneous group'. These households differ with respect to the depth of poverty they experience, the material, human and social resources they can marshal, the opportunities in their surroundings and the strategies they can employ in their struggle against deprivation. This context can be and clearly viewed from the following case.

Price heterogeneity is one such factor, which constitutes a poor heterogeneous group of the poor with respect of real inequality in income. But the price heterogeneity conditions of the poor are mostly not noticed by the conventional (quantitative) surveys. These quantitative techniques normally assume that the price for one commodity with similar quality is same for all ranges of people. But, there may be some circumstances when the prices are different for different quantities though the commodity and quality are the same. Such conditions mislead a researcher in measuring actual poverty level. This limitation may be corrected if the qualitative techniques are applied by according due considerations to ground realities in which the poor live.

The existence of price heterogeneity can be uncovered by using PRA methods before making the conventional surveys. The inclusion of this variable in the survey questionnaire would help to collect more accurate and relevant information on different poverty levels. Thus, the PRA methods enhance the focus of the quantitative methods on poverty measurement.

A study by Rao and Komala (1997) using both PRA and quantitative methods, in a potter's community in the Karnataka State of India revealed the existence of price heterogeneity-led differences in real inequality among the poor. Ethnographic survey was employed by involving focused group discussions and the key informants (in PRA methods) to explore the differential price conditions of the poor. PRA tools were found highly helpful in getting the ground realities of the poor and their different poverty levels. This was the first study in a developing country to identify the differential price conditions for different income groups of the poor.

The potter community in the study location was suffering from the cash liquidity constraints because of irregular and low level of income under seasonal employment opportunities. This forced them to buy essential food commodities in small quantities. Because of the existence of price heterogeneity for different quantity levels in this locality, the poor had to pay more as the quantity purchased by them decreased. It thus caused a reasonable difference in their real income level adjusted to the prices paid by them.

This income-depended price condition and thereby higher differences in real income of the poor would have been missed if the study was conducted without apprehending the real condition of the poor. The poverty measurement in such a study would have been certainly inaccurate. It is evident from this case that the situation where price heterogeneity exists, then it needs to be assessed in depth by using appropriate PRA techniques before executing the conventional survey. By using PRA the study could measure the real income differences among the poor and the policies framed on these results would be more meaningful and would serve the interest of the poor effectively.

Case II

Relative Price Changes and Recall Bias

Government macro-economic interventions, particularly Structural Adjustment Programmes (SAP), make important changes at the macro as well as micro (village) levels. The SAP impact assessment studies mostly consider the changes at the macro level and often ignore the changes at the village and household levels. Some of the critical factors like changes in employment structure, income, expenditure pattern, etc., in the households may remain either unnoticed or noticed only in narrow sense. Such unnoticed factors may have serious implications on the livelihood of households, particularly the poor households.

One important Structural Adjustment Programme (SAP) is devaluation of domestic currency. It is usually undertaken by a country to adjust the balance of payment by changing the trade balance. It affects the prices at both national as well as village levels. The relative price changes at the macro level can be assessed without any constraints with the conventional quantitative methods. But, the assessment of relative price changes at the village level is often constrained by recall bias. The outcome of such studies is, therefore, less reliable to make policy measures. The PRA methods help to assess the changes in prices at different time periods without any constraints. These use the indigenously available sources to depict changes in prices, which make such assessments real. The PRA enables to assess the relative changes in prices at the household level.

Mwingi District in Kenya is a semi-arid area that supports the income of its households through subsistence agriculture. Some cash income is also generated through non-agricultural sources like remittances from household members employed in towns, casual labour on neighbouring farms or in local areas and from the sale of handicrafts. In drought years, these non-agricultural sources of income, together with local coping strategies, comprise the main livelihood base.

A study was undertaken by Simanowitz, 1996 in this district to explore the impact of macro economic policies, in particular the effect of currency devaluation, on the prices of imported or exported goods and also on livelihood of rural people. One of the central aims of the study was to explore the relative price changes at the household level. PRA methodologies were used to analyze the economic changes and identify the development options for augmenting livelihood strategies.

The background in which the relative price changes were explored was that over a period of two years and at both the national and local levels, there had been high inflation in prices of goods purchased by the consumer, but the risk was low in the producer price. Also, these changes were not uniform and the understanding of the relative price changes was key to study the impact of these changes on household livelihood.

Livelihood structures (e.g., goods bought and sold, reciprocal labour, casual labour for cash and kind, bride-price payments) in the area were purposely studied. The aim was to explore the relative changes in the value of these commodities and relate them to market changes both at macro and local level as indicators of the effect of the economic policies. Matrix ranking, a PRA technique, was used to catch the relative price changes between the years 1992 and 1994, when the SAPs were introduced.

Important livelihood items of the district were identified through interviews with the local people. Since the focus was on livelihood rather than income, these included non-cash items such as labour entitlements and bride wealth. Informants, individuals and groups, were approached to assess the income and expenditure. The items for income and expenditure were indicated on one side of the matrix with 'dates' at the top. Informants used stones to indicate the relative costs of the commodities and incomes.

The results revealed that in all the cases, the decline in income relative to expenditure was more. The apparent price changes were also observed. The use of different sizes of stone enabled them to realize areas where their costs had increased disproportionately to their income, and what their income could buy in the previous years and in the current market conditions. The analysis also helped them to identify 'best' source of income, in terms of their relative value and possible changes in income generation.

Case III

Social Mobility and Poverty: Complementary Use of PRA

The social mobility is an important factor which makes large difference in access to resources and status of the individuals especially the poor households. The level of mobility is less among poor households compared to that in non-poor. It creates differences in accessing public resources and enhancing income and wealth of the individual household. The use of quantitative methods partially explains the causes of the lack of social mobility and resulting differences in the household income. An empirical study conducted in Uttar Pradesh and Bihar States in India by Kozel (1999) used both the Quantitative and Qualitative Methods to make an in-depth analysis of the access and uses of basic services and the level of their influence on the economic benefits to poor and non-poor households.

The study was based on the information generated from two sources—a conventional household and community survey and qualitative methods from Participatory Rural Appraisal (PRA). The main objective was to exploit the unique potentialities of both quantitative and qualitative research methods and to take the advantages of the possibilities for the mutual enrichment offered through the use of both methods together. The findings from this study have supported the view that combining both quantitative and qualitative methods measure the poverty and related issues more precisely.

The study found that poor, usually interpreted based on the quantitative results on the socio-economic conditions, would not be a complete measure of poverty. This is because, the poor households vary on account of other factors such as household composition, access to resources and perceived and actual vulnerability to further risk on the livelihood insecurity of the region.

The poor households are generally confined to one of the three profiles. The poorest, best interpreted as the "destitute poor," usually suffer from the idiosyncratic shock such as death of a bread earner. They may also be deprived due to lack of strong social ties, lack of physical assets and resulted economic self-insufficiency. The "structural poor", the second order in the profile of poor households tend to be the scheduled castes or tribal groups, usually suffer due to the social discrimination and live in hamlets, which are isolated from other forward community dwellings. This renders them less accessible to even essential needs such as drinking water from wells, irrigation sources, and food sources, which are common public utilities and should be available to all. The mobility (either upward or downward) among this group is also less compared to the other type of poor households. The third group can be described as the "mobile poor", who could demonstrate the strong potential of moving out of poverty.

Everyone in these three groups is confronted with a different set of constraints and opportunities for subsistence and mobility. The survey-based analysis and results showed that the poor tended to rely on the earnings from casual labour in agriculture within the boundaries of the village, whereas the relatively better-off households had diversified employment opportunities which included the lucrative employment from non-farm sectors, besides agricultural labour earnings. This difference was evident despite the equal level of education and physical strength between the two groups.

The results from the qualitative analysis (PRA) added insight into the factors that determine the access to off-farm employment and the existing local labour markets. The networks, mutual support and assistance were the factors that made difference among the different groups, despite its existence in poor and non-poor alike. It showed that the networks of the poor (especially the 'structural poor') seldom extended beyond the boundaries of the village. The relatively better-off, households could often maintain mutual assistance networks within the members of same group and with other groups in urban or government institutions which help them to secure livelihood opportunities or to appropriate a lion share in the public resources. Thus, the lack of 'social capital' was found to often deprive the poor in moving out of poverty level.

Participatory Poverty Monitoring System

The above discussions with empirical cases support the view of inclusion of PRA in the PPA. However, sustainable and systematic use of PRA warrants institutionalization of this concept in the form of Participatory Poverty Monitoring System with well-defined institutional arrangement at macro, meso and micro levels and their integration. The conceptual model of institutional arrangements with structure and functions of institutions are described in Table 4.2 and depicted in Figure 4.2.

Strengths and Weaknesses of PRA

Objective evaluation of PRA *per se* in terms of strengths and weaknesses is very important for judicious use of PRA in Participatory Poverty Appraisal. This information also helps in selecting appropriate (specific) PRA technique.

Strengths

- The PRA techniques provide an opportunity to examine the variables or issues from core of their existence by offering model environment and increasing the scope of predicting the intervening variables in the issue under investigation.

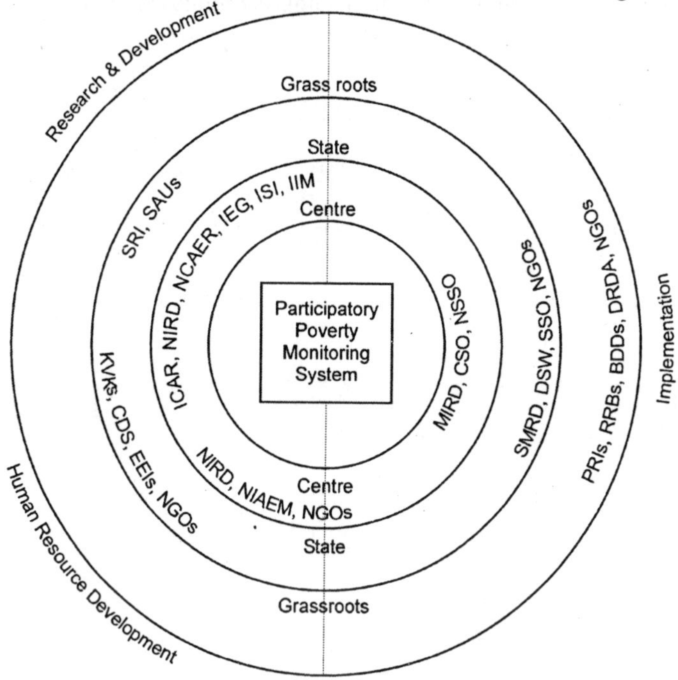

Fig. 4.2: Participatory Poverty Monitoring System : A Schematic Representation

- PRA methods are interactive in nature and facilitate maximum involvement of participants and focus their attention very close to the issues to be examined and thus provides opportunity to unearth actual factors/variables, which explain or measure the issues of interest.
- The preparatory process before the application of PRA which are essential for the execution of various methods in their relevant context, require rapport building, and credibility establishment with the participants. It helps to build confidence of the participants on enumerators and making them comfortable in disclosing the actual facts and figures attached to the issues. Moreover, it facilitates an open discussion on the cause and effect factors.

Table 4.2: Participatory Poverty Monitoring System : A Description

I. Research and Development	
Institutions	*Functions*
(a) Centre	
Indian Council of Agricultural Research (ICAR), National Institute of Rural Development (NIRD), National Council of Applied Economic Research (NCAER), Institute of Economic Growth (IEG), Indian Agricultural Research Institute (IARI), Jawaharlal Nehru University (JNU), Indira Gandhi Institute of Development Research (IGIDR), Indian Statistical Institute (ISI), Indian Institute of Management (IIM), Indian Agricultural Statistical Research Institute (IASRI).	Facilitating Participatory Poverty Monitoring System (PPMS) to frame and implement policy decisions on major issues relating to poverty like food, nutrition, livelihood infrastructure, etc. Evolving short-term and long-term poverty monitoring strategies/systems for the state as a whole and regions/districts in specific. Finding and suggesting the different ways and means of orientation of various development work in line with the policies of the government concerning poverty assessment and monitoring. Providing solutions to the issues emerging from various state and even grass roots level institutions. Conducting research on the various participatory research methodologies to evolve and test the potentialities of new methods. Further, refining the methodology of PPA for effective poverty measurement in conjunction with quantitative techniques.

(*Contd.*)

Table 4.2 (*Contd.*)

Institutions	*Functions*
(b) State	
State Research Institutions (Economic, Poverty, Agricultural and Rural Development Institutions), State Agricultural Universities (SAUs), other institutions, e.g., Madras Institute of Development Studies (MIDS) (Tamil Nadu), MS Swaminathan Research Foundation (MSSRF) (Tamil Nadu), Centre for Development Studies (CDS) (Kerala), Institute of Social and Economic Change (ISEC), (Karnataka)	Like the Centre, State R&D institutes also facilitate the Participatory Poverty Monitoring System (PPMS) at the State level. Preparing action plan for the State and grass roots level as a whole based on the recommendations of the State level agencies and the proposals received from grass roots level institutions. Prioritizing needs of particular poverty monitoring programme of State and grass roots level. Conducting location-specific research on PPM to evolve methodologies and identify potential indigenous indicators. Facilitating forward and backward linkages with grass roots and centre level institutions for a better coordination. Working out the mechanism for better co-ordination among development departments at the grass roots level.
II. Human Resource Development	
(a) Centre	
National Institute of Rural Development (NIRD), National Institute of Agricultural Extension Management (MANAGE), Non-Governmental Organizations (NGOs)	Framing the curriculum for capacity-building programmes based on the requirements of training (knowledge and skills) involved in PPM; Conducting job orientation training programmes on PRA and participating poverty assessment to various functionaries of Centre and State level institutions;

(*Contd.*)

Table 4.2 (*Contd.*)

Institutions	*Functions*
	Coordinating the training institutes at the State level and facilitating their training programmes by offering guidance and joint participation; and Assessing the training needs to meet out emerging challenges in PPM.
(b) State	
State Agricultural Universities (SAUs), Centre for Development Studies (CDS), Trainers Training Centres (TTC), Extension Education Institutes (EEI), NGOs	Organizing training programmes on participatory poverty measurement methodologies and field experience trainings to State, and grass roots level functionaries; Facilitating backward and forward linkages for joint centre and grass root level participation in preparing courses and plans, and training activities; Developing infrastructure facilities for State as well as grass root level training programmes.
III. Implementation	
(a) Centre	
Ministry of Rural Development, Central Statistical Organizations, National Sample Survey Organization	Assessing the overall requirement of resources and working out the modalities of making them available to institutions under the umbrella of participatory poverty assessment systems Conducting national level participatory poverty measurement sample surveys and creating and managing poverty databases; and Providing feedback to the national research institutes from time to time on problems emerging from poverty monitoring surveys.

(*Contd.*)

Table 4.2 (*Contd.*)

(b) State	
Institutions	*Functions*
Ministry of Rural Development, Department of Social Welfare, State Statistical Organizations (CSOs, NGOs)	Assessing the overall requirements of resources for the implementation of PPA at various levels within the State. Working out the resource (schemes) allocation patterns for grass roots in line with their requirements; Conducting participatory poverty measurement surveys and creating and managing poverty databases; Providing guidance to the grass roots functionaries on implementing the particular poverty measurement survey; and Orienting the overall PPA (Participatory Poverty Assessment) in line with the policies of the State government on poverty.
c) Grass root Level	
Panchayati Raj Institutions, NGOs, Rural Banks, Block Development Departments, District Rural Development Agency.	Preparing action plans for implementation of programmes of participatory poverty assessment at grass roots level (district/villages) based on the recommendations of the centre, and State level institutions; Forwarding information periodically on poverty-related issues to the concerned Centre and State level institutions; Familiarizing the grass roots level workers with the schemes/priorities of all line departments;

(*Contd.*)

Table 4.2 (*Contd.*)

	Assessing, mobilizing and making available necessary resources to the grass roots level functionaries; Documenting and analyzing the problems encountered during the implementation of PPM programmes and communicate these problems to the Centre and State R&D institutions; and Assessing the progress of the poverty monitoring process within their ambit and sending repose to the concerned State authorities.

- The mapping techniques in PRA methods add spatial dimensions in the problem analysis and facilitate the enquiry or decision-making process in line with the geographical conditions of locality/study area.
- The use of localized materials in the execution of most PRA methods adds strength to judgments or measurement and decisions by conceptualizing the events or variables as specific as its existence.
- Group participation, which is often used in most of the PRA approaches, facilitate the mutual understanding of the issues and also provides scope for considering the conflicts and common interests of the group in determining the issues.
- PRA has exclusive strength in identifying and explaining indigenous variables, which have unique potential in capturing location-specific indicators.

Weaknesses

- In the execution of PRA tools, "rapport building" is inevitable which warrants public relation skills on the part of enumerator. Moreover, some of the PRA tools require participation of respondents as a group. So, it becomes a tedious and resource-intensive exercise.
- In PRA methods, the data generated is mostly qualitative in nature. So, it requires appropriate treatments and transformations before a standard analysis, which is time-consuming, especially in the case of advance analysis.
- Data sets based on indigenous variables in PRA puts major restrictions in pooling data/generalization. Moreover, few statistical software packages are available to handle data collected through various PRA tools.

Conclusions

The poor constitute a heterogeneous group diversified in terms of resource endowment, opportunities, livelihood, etc. In this mode, no single measurement technique can cover the entire sphere of poverty and thus warrants integration of different measuring techniques. Most of the poverty-related or proxy indicators are of indigenous in nature and have important role in defining poverty in its real sense. These indigenous indicators often get escaped from the eyes of conventional measuring techniques. PPA, which has the advantage of, combined use of both qualitative and quantitative methods, takes care of those indigenous factors which influence the poverty. Participatory Poverty Appraisal offers a wider scope for using various PRA methods. These methods are diversified in nature to handle the different conditions in which poverty exists. Moreover, these methods can be combined with other conventional techniques and increase the scope of measuring the dynamic poverty indicators. However, systematic methodology is to be adopted for judicious use of PRA techniques and combining it with other quantitative techniques. For a wider application of the combined use of qualitative and quantitative tools, efficient participatory poverty monitoring system is to be introduced at various levels. The system should be comprehensive and their units should function in integrated way both vertically and horizontally. Further knowledge of strengths and weaknesses of PRA would enable judicious use of PRA, leading to precise Participatory Poverty Appraisal. Such a scientific and comprehensive approach can help better understand factors causing poverty and make listening to voices of the poor better.

REFERENCES

Barret, Chris. 2001. Integrating qualitative and quantitative Approaches: Lesson from the Pastoral Risk Management Project, In. *Qualitative and Quantitative Poverty Appraisal*: Complementarities, Tensions and the Way Forward, Workshop Proceedings, Cornell University, March 15-16.

Haan, A.D. and J. Koch Laier. 1997. Employment and poverty monitoring, In. Issues in Development, *Discussion Paper No. 19*, ILO, Geneva.

Hadi, A. and A.M.R. Chowdhury. 2001. Poverty monitoring and evaluation in BRAC, Bangladesh, *http://www.worldbank.org*.

Kozel, Valerie, 1999. WDR on poverty and development 2000/01, Stiglitz Summer Research Workshop on Poverty, Washington DC.

Kozel, V. and B. Parker. 2000. Integrated approaches to poverty assessment in India, In. M. Bamberger, (Ed.) *Integrating Quantitative and Qualitative Research in Development Projects*, Washington, DC, World Bank.

Ravallion, M. 2002. How can quantitative methods help in measuring poverty? In., *Economic and Political Weekly, Special issue on Poverty*

Measurement, Monitoring and Evaluation, 38(4): 296-301.

Rao, Vijayendra and A.C. Komala. 1997. Are prices higher for the poor? Price heterogeneity and 'Real' inequality in rural Karnataka, *Economic and Political Weekly*, 38(3).

Simanowitz, A. 1996. Analyzing the Impact of Macro-economic Policies using PRA in Muingi District, Kenya, PLA Note, and No. 26.

5

Participatory Evaluation and Impact Assessment : A Case Study

M.J. Chandre Gowda and *S.H. Adapur*

Participatory methods provide active involvement in decision-making for those within a project, programme, or strategy and generate a sense of ownership in the M&E results. These methods are used for learning about local conditions and local people's perspectives and priorities to design more responsive and sustainable interventions. Besides identifying problems and trouble-shooting problems during implementation, the participatory methods help in evaluating a project, programme, or policy and providing knowledge and skills to empower poor people. *Participatory monitoring and evaluation,* as one of the participatory methods, involves stakeholders at different levels working together to identify problems, collect and analyze information, and generate recommendations.

Other commonly used Participatory Tools/Methods are Stakeholder analysis, Beneficiary assessment and Participatory Rural Appraisal.

Stakeholder analysis is the starting point of most participatory work and social assessments. It is used to develop understanding of the power relationships, influence, and interests of the various people involved in and activity and to determine who should participate, and when.

Beneficiary assessment involves systematic consultation with project beneficiaries and other stakeholders to identify and design development initiatives, signal constraints to participation, and provide feedback to improve services and activities.

Participatory Rural Appraisal (PRA) is a planning approach focused on sharing learning between local people, both urban and rural, and outsiders. It enables development managers and local people to assess and plan appropriate interventions collaboratively often using local people. The PRA tools are often only seen as appropriate for gathering information at the beginning of an intervention and as part of a process of appraisal and planning. Development workers may talk about having 'done' a PRA, sometimes seeing it as just a step towards getting funding. However, PRA tools have a much wider range of potential uses, and can often be readily adapted and used for participatory monitoring, and for participatory evaluation.

A Case Study of using PRA Tools

Farm Science Centre (popularly known as Krishi Vigyan Kendra or KVK) is the district level organization for promotion of scientific agriculture among farming community. Vocational training is one of the mandates of the KVK aimed at creating additional livelihood opportunities for the rural families, both land-owning and landless. Krishi Vigyan Kendra in Gadag district of Karnataka State has been working through Group Approach for greater effect of its activities. The activities promoted through one of the group has been studied and analyzed to demonstrate the utility of participatory impact analysis procedures. The KVK worked with this group since 1999. Formation of group, awareness building on the group procedures, skill building on group functioning and capacity building on technical activities were supported by the KVK.

To support the adoption of the technologies which were imparted through training and exposure visits, the KVK provided critical inputs in the form of seeds and planting material produced in KVK farm. Some other inputs like buffaloes, feed and fodder material were procured by the members from the bank loan or from the group corpus fund. In addition, the KVK scientists visited the group at critical stages of various enterprises to support the technology utilization.

Details of the Trainings Imparted to the Group

1. **Goat Rearing**
   - *Feed and fodder management* : Cultivation of green fodder, Dry fodder enrichment, Preparation of concentrated feed, etc.

- *Disease and pest management* PPR management, Mastitis management, Rinder-Pest management, Vaccination, Deworming against ecto and endo parasites
- Insurance against mortality

2. Buffalo Rearing

- *Balanced nutrition* : Feeding of green fodder, Enrichment of dry fodder, Preparation and feeding of concentrate feeds
- *Cattle shed management* : Proper drainage system, Floor space for milking animal, Proper hygiene and ventilation, Pest management
- *Disease and pest management* : Hemorrhagic Septicemia, Black quarter, FMD and Mastitis management, Vaccination and First Aid
- Care and management including Milking methods

3. Kitchen/Nutrition Gardening

- *Nursery management* : Placement of seeds/seedling, Raised bed, Pest management, Staking of vegetables
- *Cultivation aspects of vegetables* : Choice of crops, time of sowing, Sowing techniques, Seed treatment, Irrigation methods, Pot/drum method of irrigation, Pest and disease management
- *Importance of human nutrition* : Importance of fruits and vegetables in daily diet, Hygiene practices for fruits and vegetables

---

Three PRA tools used in analyzing the impact of interventions of a Krishi Vigyan Kendra on the standard of living of members of a SHG have been described here. The facilitators, first introduced the purpose of the evaluation exercise and the steps involved in each method/tool.

1. Time Line

A time line enables the stakeholders/participants to understand the history of the village, community or programme through the chronology of developments, marked by landmark events that many people can recall. It can as well be used to compare 'before and after' situation, as is done in the present case.

Steps Followed

1. Explain the subject of focus to the participants and decide how it will be represented physically (written words, drawing, etc.).
2. Set the period of time line (1999 to 2006, in the present case)

and explain to the participants on the reasons behind considering a particular period.

3. Facilitate participation of all members and encourage discussion. Ask clarifying questions and details of events as and when required. Focus should always be on the issue being analyzed.
4. Make sure that the participants are comfortable with the medium used and include representation with objects or through the use of drawings.
5. Analyze the results/findings with the members about the related problems and opportunities.

The time line of activities conducted by the KVK are given below :

Year	*Major Activities*
1999	• Organized into SHG, capacity building on SHG concept (2 days)
2000	• First loan linkage for consumption purpose
2001	• Training on Goat rearing (2 days) and Kitchen Garden (1 day)
2002	• Training on buffalo rearing for 3 days, enterprise initiated with bank loan
2003	• Awareness training on backyard poultry
2004	• SHG linked to National Agriculture Technology Project (NATP) programme for food processing
	• Awareness on food processing activity (vermicelli enterprise)
2005	• Exposure visit to study the food processing activities
	• Vermicelli unit started (Group activity)
2006	• Participation in Wednesday Bazaar for sale of products, every Wednesday from 4 pm to 8 pm.

Impact of the Changed Livelihood Options on the Food Consumption Pattern

The matrix given below reveals that rural households preferred to lead a lifestyle similar to that of their urban counterparts. Over the period of time there has been a decrease in consumption of pulses and increase in consumption of vegetables. It may be due to easy availability of vegetables from the kitchen gardens which the group members started after undergoing training. Increase in price of pulses may be the reason for less consumption of pulses. Jowar consumption was reduced mainly due to lack of sufficient time for preparation of *rotis* and also due to availability of rice at cheaper rate through Government supported Public Distribution System.

The combined tool of matrix ranking and time-line was used for comparing the dietary pattern before interventions by the KVK (2000) and at the time of impact analysis (2006). The KVK helped the participants with the help of pictures of vegetables and fruits to use them as identification tools. These were lined up on the middle of a sheet, and the members placed stones to indicate the quantity of vegetables used 'before' on the 'left side' of the pictures and 'after' on the 'right side'.

Comparing the dietary pattern before and at the time of evaluation

Particulars	*2000*	*2006*
Quality of Vegetables	OOO	OOOOO
Types of Vegetables	OOOOOOO	OOOOOOO
Quantity of Vegetables consumed per month	OOOO	OOOOOO OOOOOO
Quantity of Pulses consumed per month	OOO	OO
Quantity of Sorghum (*Jowar*) consumed per month	OOOOOO	OOOOOO
Quantity of Rice consumed per month	O	OOO
Monthly visits to Hospital	OOOO	OO
Hospital Expenditure (Rs./visit)	OOOOO OOOOO	OOOOO OO
Productive working days/week	OOOO	OOOOOO

There was sufficient saving on the cost of vegetables due to increased availability of fresh quality vegetables. Women also saved time on the purchase of vegetables from the town.

Number of visits made to hospital for treatment of family members reduced drastically from about four visits a month to just two. Reduced illness, nutritive food consumption and education on hygienic living conditions have led to this change. Number of days of productive work was doubled.

They listed the following major changes in their life due to the KVK interventions :

- Change in livelihood options,
- Increased employment,
- Reduced dependency on money-lender,
- Improved nutrition and health, and
- Increased self-respect, respect within the family and from the community

Change in Livelihood Options

Livelihood options of woman members of the group increased manifold. These were the actual sources of income for the group members. There were four additional sources of income to the earlier one namely, wage work. The added sources of income were kitchen garden, goat rearing and buffalo rearing as individual activities and the Vermicelli unit as the group activity (Fig. 5.1).

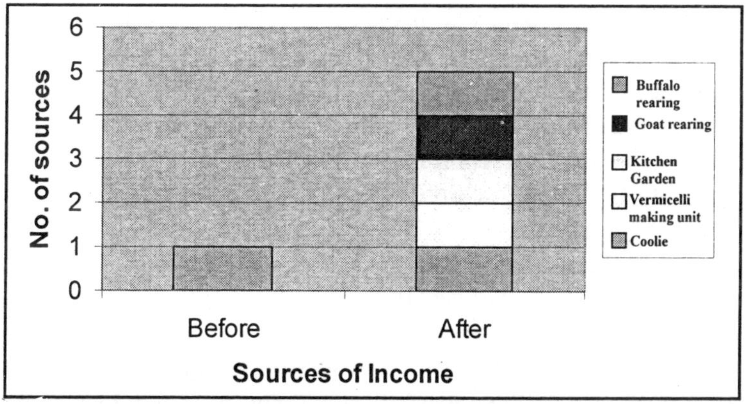

Fig 5.1: Chart showing increased livelihood options for women

Increased Employment

The direct impact of increased number of livelihood options was that the employment of members doubled, from 6 hours a day (only seasonal) to 13 hours a day (4 hours regular and 9 hours seasonal). While the 6-hour seasonal wage work earned them rupees 30 per day, the increased livelihood options enhanced their daily income to rupees 150 per day, out of which rupees 60 per day was assured throughout the year (Fig. 5.2).

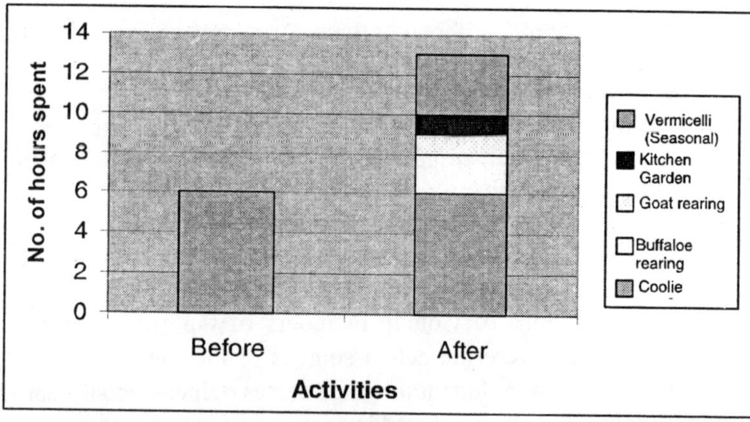

Fig. 5.2: Chart showing working hour status of women

Reduced Dependency on Moneylender

The immediate outcome of the increased income and the livelihood options was that the dependency on moneylender stopped. Assured income and the source of loan either from banks and/or from the group corpus fund enabled them to come out of the clutches of moneylenders. The group corpus fund was handy for the home consumption loan, which was not easily available from the commercial banks.

Improved Nutrition and Health

Consumption of vegetables and fruits by the members' families increased by six times after the intervention of kitchen garden activities. New vegetables like carrot, beetroot and tomato were added to their daily diet. Quantity of onions, okra, and brinjal consumed also increased. Fruits became part of their diet for the first time. Inclusion of guava, sapota, lemon and papaya were evident in the increased nutrition of the members.

Increased Self-Respect, Respect within the Family and from the Community

Exposure to the SHG activities increased to confidence level of women members. It was evident not only in their dealings with the neighbourhood, or with the members of the opposite gender in the community, but also with the various agencies and individuals of the outside world. This was mainly due to the confidence in them that they, as individuals and members of the groups, possess abilities to perform better. Abilities were related to technical, behavioural, economical and social. It was evident that the members were in a new world of opportunities, and a world of empowerment. Attitude of husband/male members within the family changed for better.

2. The Spider Web Diagram

Evaluating Changes due to Programme Interventions using Spider Web Diagram

The spider web diagram is also called a cobweb diagram, participation wheel or an evaluation wheel. It is a highly visual method for analyzing the relative importance of, or progress on, different aspects of an intervention. This exercise can be done to plan projects, but particularly to monitor and evaluate them. Each aspect is represented by one arm of the frame of the web, and is graded from 1-10. It is also possible to rank programme/village/group/individual performance during (monitoring) or at the end (evaluation) of a programme. The spider web diagram is a relatively quick and easy tool for progressive impact assessment, which participants can readily observe and feel.

Steps in conducting a Spider Web Diagram

1. Explain the purpose of the exercise to the participants, and reach a common decision on the aspects to be rated by them, depending on the exercise. These could be the performance of the project against the original objectives, the extent of developments in different villages, etc.
2. Encourage the participants to write the aspects to be rated on cards or represent them by drawings. Take a large paper

and arrange the cards in a radial form away from the centre. Join the centre to the card with a straight line (See diagram). This line represents a scale from 1-10.

3. Facilitate the participants to arrive at a consensus score on each of the aspects based on the performance and help them to mark the score on the axes of the circle. This can be done after thorough discussion to reach consensus on a collective score.
4. Join the score on each of the axes, with a line as shown in the diagram.
5. Discuss the scores and the underlying causes/reasons.
6. Document the exercise and take note of the reasons stated by the participants. These results can then be used for participatory planning, monitoring and evaluation.

In the present case, the group members were most comfortable with comparative rating, rather than assigning specific numerical values (scoring). The important thing was to ensure that everyone understood and were at ease with the methods used. As it was a small group of 10 members, dominance by a few individuals was not noticed.

The spider web described here demonstrates the significance or the intensity of contribution of KVK to the life of the group members. The members chose six indicators to highlight the qualitative changes that have taken place among the members :

(i) Ability to spend in emergency;
(ii) Household articles possession;
(iii) Possession of productive assets;
(iv) Status of house;
(v) Recognition of group;
(vi) Recognition of individuals as resource persons;

Some of the criteria identified by the participants were unconventional and surprising. To indicate the criteria of self-respect, the members used their group's rubber-stamp, and to indicate their reduced dependency on moneylender, the members displayed their bank pass-book.

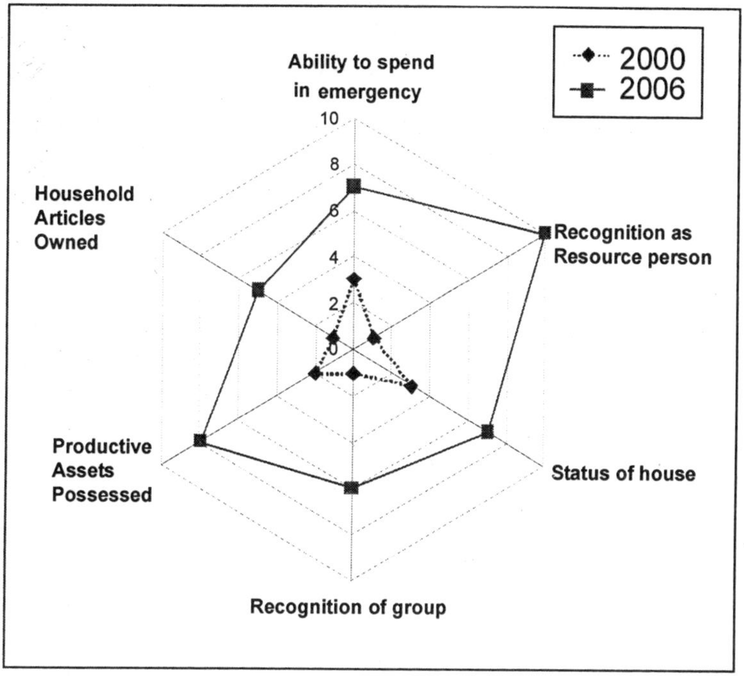

Fig. 5.3: Spider Web depicting the changing Socio-Economic Status of the Women

As individuals, the members felt that maximum change happened in the form of "Recognition as Resource Persons" in their areas of work like livestock management, kitchen nutrition gardening and group management. The members had no hesitation in giving maximum of '10' score, as compared to hardly any recognition level previously. The change in 'productive assets possessed' also had substantial increase from score of '2' to '8' in a span of six years. The group felt that there is lot of scope for improvement in the status of 'household articles possessed', 'recognition for the group' as a whole and their 'status of houses'.

The web diagram indicates three dimensions of the life of the group members. The inner-most circle indicates their status at the beginning of KVK intervention in the year 2000. The middle portion of the web indicates the quantum of change in the lives of the members due to the intervention of the KVK. The outer space in the web diagram indicates the potential or the distance to go to achieve the intended change in the life of individual member or group as a whole. This way, the spider web method of impact evaluation can be used to assess the progress of any development programme 'at the beginning', and 'at desired intervals'. The gap between the target or the programme objectives and the actual achievements at various stages are also clearly elicited in the spider web.

The web diagram clearly demonstrates the increased gains for the group both individually and collectively as the direct impact of interventions carried out by the KVK. It also indicates the areas for further improvement.

Conclusion

The case study has brought out the changes on the life of very poor people due to the efforts of KVK in a short span of time. Series of interventions, covering multiple activities, each opening-up a new/additional livelihood option to the group members have resulted in social and economic impacts. It is true that quantification process of the changes that have taken place due to the intervention of KVK could have also been done by collecting data through interviews, observations and or feedback. However, participatory approaches described here provided opportunity for learning for all the stakeholders (The financing and coordinating agency—Indian Council of Agriculture Research (ICAR), implementing agency—KVK, and the beneficiaries—the group members). The group members could see from their own eyes the significant changes that have taken place in their day-to-day life which were clearly visible through pictures, graphs, etc. The participatory learning also enabled the KVK to realize that changes happen over a period of time and the cumulative effect gives rise to greater impact. The indicators used, the method of quantification chosen by the participants, and the confidence with which they carried out the impact assessment proved that the participatory monitoring and evaluation tools can be easily adopted and practiced. Results that emerge from such an analysis are verifiable and hence pass the reliability and validity requirements.

6

Participatory Impact Monitoring and Assessment

Vidya Ramachandran and *M.J. Chandre Gowda*

Despite the increasing number and sophistication of management tools and methodologies, monitoring the impacts of development efforts continues to be a complex and neglected task. Management focus is generally concentrated more on planning than on other aspects of project administration. Results are usually measured in terms of outputs and sometimes in terms of outcomes, but almost never in terms of impacts.

Why Impacts are Rarely Measured?

Impacts are often difficult to measure because of some of the following reasons :

- Impacts do not always happen as per plans and schedules,
- Impacts that are intangible or qualitative are difficult to measure and document credibly and comprehensively,
- Unintended, unplanned, unexpected impacts get overlooked unless they are specifically discovered and captured,
- The extent to which project activities alone are responsible for impacts is not always clear since there may also be other external factors influencing impacts, and
- Practical methodologies to assess and document impacts are inadequate.

On the other hand, development agencies are increasingly exposed to public pressure and are expected to justify how and to what extent expenditures have benefited the intended populations. They are called upon to demonstrate that their projects are creating the expected benefits for their target groups.

Participatory Impact Monitoring/Assessment refers to a process in which development interveners and local communities jointly observe, document and critically reflect on the effects and changes caused by project interventions.

Objectives of PIM/A

The three-fold objectives of PIM/A are :

- Promoting Learning Process;
- Improving Communication between Stakeholders, and
- Improving Project Steering.

PIM/A is not simply a methodology but it represents a philosophy. It is not a one-time event, it has to be periodically undertaken so that programmes and intervention strategies are constantly reviewed and improved.

Why PIM/A useful in Development?

- Change in the context of development is usually the result of a process;
- This process is made up of many contributing pieces of action, with each piece of action leading to another stage in the change process;
- Often in such a process, the intermediary changes get noticed because they happen more immediately, are often more visible and cause more excitement;
- In doing so, the ultimately desired change can get missed or forgotten or it may even simply be assumed to have occurred because the intermediary signs were good; and
- PIM/A ensures that the desired change to be monitored/ assessed is not lost sight of.

Pre-conditions for PIM/A

Assessing impacts of a project's actions is essentially a management requirement to provide the basis for improved project screening. It enables identification of 'weak' areas and 'deviations' on the basis of which corrective actions can be made.

PIM/A cannot remain a one-off exercise; for best results it has to become a *systematic feature* of the organization. Thus, it needs a set of planned impacts for each programme.

- A stable project structure and longer duration project commitment;
- Good communication channels between people, including the targetted communities;
- An interest in PIM/A and a willingness to adopt it despite the extra workload entailed;
- Adequacy of resources to be invested in PIM/A (staff time, computers, transport facilities etc.);
- A work culture and community approach that is committed to participatory principles;
- Skills to facilitate the process (defining impacts, developing indicators, interviewing, analysis, documentation etc.);
- Management willingness to learn and improve, not just to implement activities. Willingness to act on the results coming out of PIM/A; and
- Harmonizing routine work and other M & E procedures with PIM/A and not executing this as an isolated exercise outside the project management cycle.

Definitions of Key Terms

Participatory = with the participation of;

- those who are interested in creating the impact. This includes people responsible for the project implementation, final users of the system, agencies funding the project, etc; and
- those who are expected to experience the impact. This would

include the people and institutions targetted for change, e.g., farmers, extension functionaries, rural youth, etc; and,

- if possible, others whose support is also important to the project, e.g., evaluators, policy-makers, collaborating institutions, etc., the involvement of this group is useful but not critical to the process.

Participation is essential :

- in defining impacts;
- in drafting indicators;
- in exploring relations between impacts, project actions, and other external influences;
- in developing data collection tools; and
- in drawing conclusions based on results.

Key Questions to be asked by the PIM/A Team

- Are all the people who really should be participating represented here?
- Is any person not there who really ought to be participating?

Participatory Monitoring and the Steps involved

It is a process of continuous observation, systematic documentation and critical reflection specially applied in relation to project impacts. The process includes :

A. Deciding on the impacts to be monitored;
B. Developing indicators to assess each impact;
C. Developing tools to collect information with regard to indicators;
D. Collection of information;
E. Analysis and interpretation of processed information;
F. Assessing the direction and quality of progress from the information collected; and

G. Using the information to feed into project management to improve project steering.

When the cycle is completed, steps D to G start all over again. The periodicity of the repeat cycles can be fixed based on the nature of programmes and the management requirements.

The steps A to C can also be reviewed at appropriate intervals to see if they are still valid or need modification.

Indicators of Impact Monitoring and Assessment

Concrete evidence or proof that can serve to confirm a claim and that can be measured or at least seen and described is called an as indicator. In the case of PIM/A, an indicator delivers a detailed description of an impact in order to assess it. Developing convincing indicators is one of the most challenging aspects of PIM/A. As projects move from technical, economic and environmental spheres (visible, quantifiable) to political, social and cultural spheres (intangible, difficult to qualify or quantify) indicators also move from being more objectively perceivable to becoming more and more subjectively determined.

Attribution—Establishing the Cause and Effect Relationship

It refers to the extent to which changes in any situation can be assumed to have occurred *because of Project Actions*. It is essential to be assessed because, change may be caused not only by the one who desires it; and it may also be influenced by the actions of other factors in the environment. Though, some external influences may be helpful in achieving the desired change, some may hinder its achievement. It is a fact that these external influences often cannot be controlled by the Project. Nevertheless, they have to be considered to determine the true influence of the Project. Development projects operate amidst complex sets of factors. As more factors come into play, the problems of attribution increase due to an aggregation of factors influencing the change.

Example

Lower level Aggregation	*Middle level Aggregation*	*Higher level Aggregation*
↓	↓	↓
Milk Production Increases	*Milk Economy Improves*	*Living Standards Improve*
↓	↓	↓
This may be attributed to a relatively smaller set of factors like more animals, better animals, better fodder, and improved veterinary care. In what areas and to what extent has this been the result of the Project's actions?	The causal factors increase. In addition to production factors they may also include better storage, better transport, and access to new markets. To what extent has this been the result of the Project's actions?	The causal factors become much more complex. Is it due to milk production? Is it due to supplementary activities surrounding the milk economy? Is it also because of other new livelihood opportunities that have developed in the area? In what areas and to what extent has this been the result of the Project's action?

Types of Impacts

Impacts of project interventions can be of three types. They are:

(i) *Intended Impacts :* i.e. impacts that are desired and planned for;
(ii) *Unintended Impacts :* i.e. impacts that are not specifically planned for but they may still be expected or anticipated; and
(iii) *Unexpected Impacts:* i.e., impacts that are not anticipated but come as a complete surprise

Example

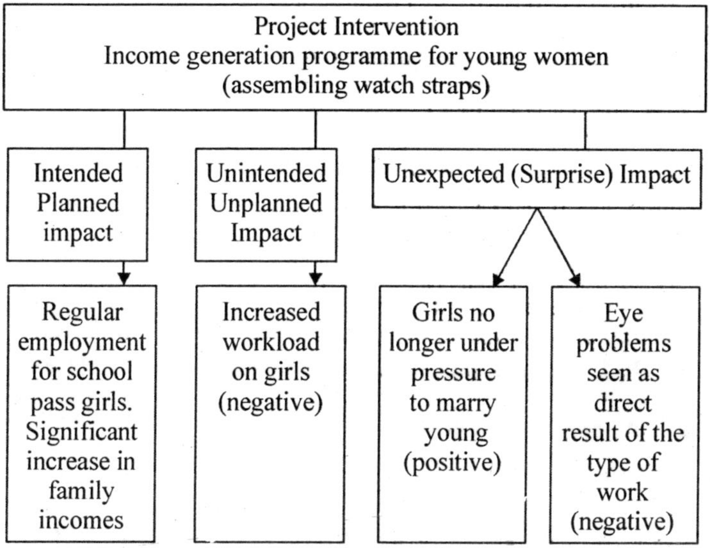

A good PIM/A encourages reflection on all types of impacts, including unexpected impacts as they begin to be noticed.

Steps in PIM/A

Prerequisite Steps

1. An organization-level orientation to PIM/A; and
2. Formation of training of a PIM/A Team at Organization/ Project level.

PIM/A Implementation Steps

1. Deciding on the programmes to be monitored/assessed;
2. Identifying the (intended and unintended) impacts of each programme selected for assessment;
3. Deciding on which of the above impacts to monitor/assess;
4. Defining key terms so that there is a clear shared meaning of the impacts;
5. Investigating the relationship between project activities and the impacts selected for monitoring/assessment;

6. Investigating the relationship between other (external) factors and the impacts selected for monitoring/assessment;
7. Drafting appropriate indicators to measure each selected impact;
8. Developing data collection tools and methods for data collection;
9. Defining survey units and deciding on sampling procedures;
10. Collecting information and processing the information collected;
11. Analyzing and interpreting the results; and
12. Drawing conclusions.

In order to internalize the steps involved in PIMA, it is essential that the PIMA team answers the following questions for each step :

- *Rationale:* Why is this step necessary?
- *Outputs:* What is the result expected from conducting this step?
- *Actors Involved:* Who should be involved in conducting this step?
- *Procedure:* What is the best possible way to conduct this step so that the required outputs are realized?
- *Checking Questions:* To confirm that the outputs of this step have been achieved

Questions to be answered in some of the steps of* PIM/A *are detailed below:

(i) Deciding on the Programmes to be Monitored/Assessed

Rationale : Development projects usually consist of more than just one type of programme. It may not be possible for all programmes to be monitored equally intensively. Therefore, choices have to be made. These choices are most often based on what are seen to be core programmes or vitally important programmes.

Outputs : A consensus is reached on the question of choices, and a manageable selection is made.

Actors Involved : Members of the organization/project opting for PIM/A. Who among them has to be a management decision? If a

PIM/A Team is already formed and trained, then that team has to be definitely involved.

Procedure : A brain storming approach can be adopted to list down all the programmes that the project is engaged in. The programmes can then be sorted according to importance, budget allocation, outreach, felt need for improving steering, present lack of awareness on impacts, strategic importance from a macro perspective, etc. A discussion on the outputs and matching it with the resources available for PIM/A will enable making the final choice.

Checking Questions : Has a manageable choice of programmes been made? Is everyone comfortable with this choice?

The assessments are subjective but valid since they are from the perspective of the organization involved and interested in PIM/A. Self-help groups, watershed development, and training score high on many counts. If time and resources permit, they can all be included. Otherwise, a further choice will have to be made.

(ii) Selecting the Impacts to be Monitored/Assessed

Rationale : Project interventions are made with the anticipation that they will have favourable impacts. These impacts have to be named if they have to be monitored. Each programme generally has several impacts, some intended and some not. Impacts may also differ from one stage of the programme to another. It is not possible to monitor all the impacts. Hence, choices have to be made on inputs to be monitored.

Outputs : A Manageable Number of Impacts are selected for Monitoring/Assessment.

Actors Involved: Representatives from the project management, implementers, donors and beneficiaries. Who and how many has to be a management decision.

Procedure : Participants must be informed in advance of the date for this step so that they may reflect on it, consult with others, and come prepared with a possible list of impacts. The selection procedure can involve two steps :

(i) A brainstorming session to generate a list of all possible impacts, viz., intended, unintended, known or assumed. Two questions can help this process:

- What positive changes did we intend from this programme?
- What unintended changes (positive or negative) do we expect or fear from this programme?

(ii) The second step consists of sorting through the list, prioritizing and arriving at a manageable list of impacts (4 to 6)

- If it is the organization's first experience of PIM/A, it is better to choose short-term (intermediate) impacts that are easier to assess.

Checking Questions : Has a manageable choice of impacts been made? Is this choice compatible with the objectives of the programme? Are all the actors involved comfortable with this choice?

Example: Programme on forming and strengthening SHGs

Listing of Impacts

Intended

- Creation of norms and effective sanctions in the group;
- Increased knowledge and awareness of members;
- Increased bargaining power of SHGs/members;
- Increased decision-making power of women members in their families;
- Increased savings and incomes of SHG members;
- Leadership development among group members;
- Development of managerial skills among SHG members;
- Development/enhancement of individual skills among members or their families;
- Increased community consciousness of members;
- Increased cooperation and trust within SHGs;
- Increased linkages between SHGs and other institutions;
- Increased realization of self-potential of SHG members; and
- Increased self-confidence of group members.

Unintended

- Increased conflicts in members' families (negative)

- Increased workload of women (negative)
- Increased influential power of SHGs in the community (positive)
- Domination of some members by others (negative)
- Continued dependency of SHGs on outsiders for certain functions (negative)

(iii) Indicator Development and Selection of the Most Appropriate Indicators

Rationale : Impacts can be estimated only by means of indicators. Indicators are what make an impact observable or measurable.

Outputs : Appropriate indicators are finalized to measure each of the selected impacts

Actors Involved : Representatives from the project management, implementers, and beneficiaries. Who and how many has to be a management decision. It is important that 'professional' viewpoints of the PIM/A implementers are reconciled with the indicators that make maximum sense from the target group perspective.

Depending on the in-house level of knowledge and experience, it may be necessary to involve other experts also to develop indicators, especially for specialized areas.

Procedure : There are several steps involved. The first list of indicators is generated by asking *opening questions* like "How can we recognize that knowledge and awareness of LEISA practices has spread in the area" This list is pruned and fine-tuned by asking *clarifying questions* like "What do you mean by farmers are 'more concerned' about managing natural resources? How exactly do they demonstrate this concern?". Further fine-tuning is done by asking *checking questions* like "Can we conclude that just because a farmer is using more Farmyard Manure, in the field is a Low External Input Sustainable Agriculture (LEISA) LEISA farmer?"

Other facilitating questions at this stage are" "Why do we think this is a good indicator to describe this impact?" "How can we convince others that this is a good indicator?" 'What aspects of the impact are covered by this indicator and what are left out?"

With all this, we may still end up with a complicated set of indicators that are difficult to administer. Collecting data can become time-consuming and expensive. Hence, it is useful to organize the selected indicators into the following matrix before finalizing which one to adopt and which one to keep aside:

Criteria	*Impact Indicators*			
	Ind. 1	*Ind. 2*	*Ind. 3*	*Ind. 4*
Validity (least limitations)				
Ease with which data can be collected on the indicator				
Is the information that this indicator will provide really necessary?				
The extent to which this indicator will yield precise data				
How cheap or expensive is it to collect data on this indicator?				

Checking Questions : Have a manageable set of indicators been selected? Are they meaningful indicators? Has the process of indicator selection been understood by all?

Note : In spite of all this efforts, indicators will still carry certain limitations, more so in the social sectors. While we should keep the search going for 'perfect' indicators, we will have to work with what is practical and possible until then.

Ms. Anke Schurmann, a development professional from Germany, was one of the first persons through whom several organizations in South India got exposed to the systematic application of PIM/A in the field. The use of these tools in the context of monitoring SHGs and watershed projects has been documented in a User's Manual authored by her titled Participatory Impact Monitoring. Exposures such as these have enabled many Krishi Vigyan Kendras (KVKs) to carry forward the PIM/A process in their respective programmes.

7

Logical Framework Approach for Planning Impact Monitoring and Evaluation

N. Nagaraja and *N. Manjula*

Introduction

The Logical Framework Approach (LFA) is a framework for designing change process, monitoring progress and evaluating impact. The first logical framework was developed for United States Agency for International Development (USAID) by the end of 1960's and has since been utilized by many of the larger donor organizations. It is also called as ZOPP (Ziel Orientierte Projek Planung) by German Agency for Technical Cooperation (GTZ) and OOPP (Objectives Oriented Project Planning) in Norway. The LFA is an analytical, presentational and management tool which can help planners and managers to (a) analyze the existing situation during activity preparation, (b) establish logical hierarchy of means by which objectives will be reached, (c) identify potential risks in achieving the objectives, and to sustainable outcomes, (d) establish how outputs and outcomes might best be monitored and evaluated, (e) if desired, present a summary of the activity in a standard format, and (f) monitor and review activities during implementation.

In designing change process, monitoring progress and evaluating impact, ordering the elements of the change process in a logical structure, is very crucial. Firstly, inputs or resources have to be identified as necessary means for performing activities. Specific activities are then identified which result in outputs, and are necessary to achieve objectives. Two terms—log frame and logical framework Approach—are sometimes confused. The log frame is a document; the logical framework approach is a project design methodology.

Basic Principles

- The log frame should be concise. It should not normally be more than two sides of a paper;
- The log frame should be treated as a stand alone document and should be comprehensive to those using to it for the first time;
- If beneficiaries are included in the project, they should also take part in the design of the logical framework; and
- The log frame will provide a basis for subsequent monitoring and evaluation. It must, therefore, be kept under regular review and amended whenever the project changes occur.

Steps in LFA

There are four major steps in conducting LFA namely—(a) Situation analysis, (b) Strategy analysis, (c) Project planning matrix, and (d) Implementation.

A. Situation Analysis

Situation Analysis includes analysis of stakeholders, problems and objectives.

Stakeholder Analysis is analysis of problems, fears, interests, expectations, restrictions, potentials of all important groups, organizations and institutions, implementing agencies, other projects and individuals. The tools that can be used for stakeholder are stakeholder analysis matrix, SWOT analysis, venn diagrams and web diagrams.

Problem Analysis is the identification of the main problem and developing a Problem Tree through an analysis of cause and effects. The problem tree is developed by moving problems from the clusters of problems on the wall and by adding new problems that emerge, as the tree is developed. Problems can be moved up or down the tree as required. The tree should end up with one main problem and a series of lower order problems that branch out below the main problems. The problem tree is constructed by selecting a problem from the list and relating this problem to a starter problem using the cause-effect ratios that if the problem is a cause of the starter problem, it is placed below the starter problem. If the problem is an effect of the starter problem, it goes above. If it is neither a cause nor effect, it goes at the same level (see Fig. 7.1)

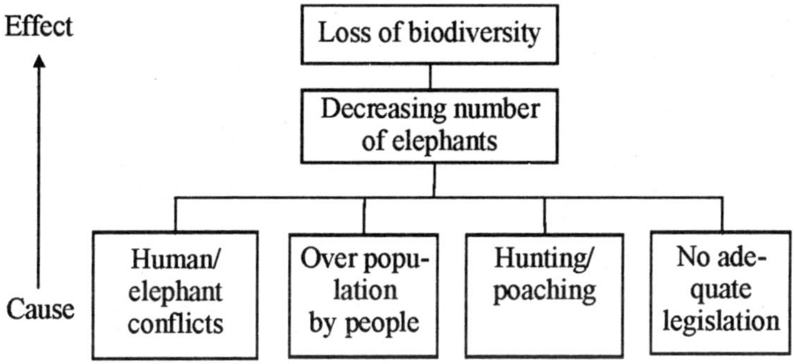

Fig. 7.1: Problem Tree

Objectives Analysis is systematically identifying, categorizing, specifying and balancing out objectives of all parties involved in a specific situation. The problem tree is transformed into an objective tree by restating the problems as objectives. The objectives tree can be viewed as the positive mirror image of the problem tree. It is also considered as ends-means diagram. The top of the tree is the end that is desired and the lower levels are the means to achieve the end (See Fig. 7.2).

B. Strategy Analysis

It is a systematic way of searching for and deciding on problem solutions. The choices among different solutions to problems may be concerning to : (a) overall concepts, strategic plans, objectives, (b) people, target groups, organizations, agencies, (c) methods, procedures, process, (d) technologies, services, products, outputs, and (e) measures, actions, materials, inputs. The strategy analysis involves clustering of objectives and examines feasibility of different

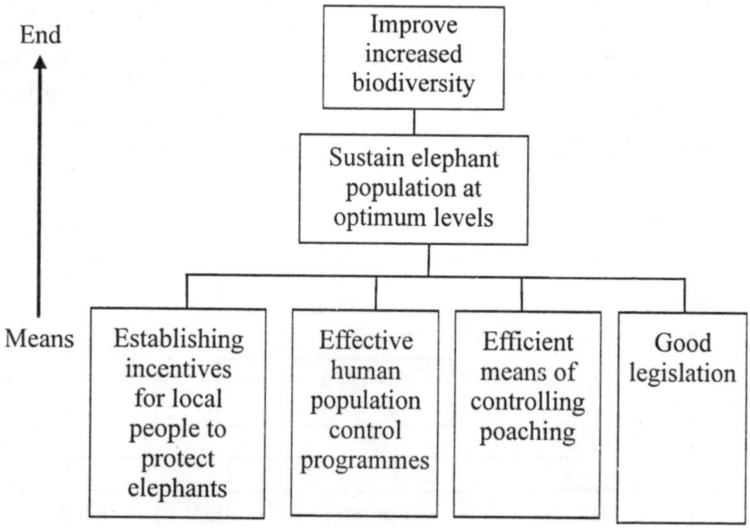

Fig. 7.2: Objective Tree

interventions. The main objective becomes the project purpose (immediate objectives) and the lower order objectives become the outputs or results and activities. The strategy analysis also looks at feasibility of different interventions. For example, effective human population control programmes may not be a feasible intervention and hence not included in the strategy analysis (Fig. 7.3)

C. Project Planning Matrix

The LFA consists of a set of interlocking concepts that facilitate planning, implementation and evaluation of a project. The LFA centres around a matrix (Project Planning Matrix—PPM), which is completed by following a standardized terminology and a series of principles. Main matrix of LFA consists of four items, viz., narrative summary, objectively verifiable indicators, means of verification and important assumptions.

(i) Narrative Summary

It defines the project structure. The components of project structure are: (a) development objectives, (b) immediate objectives, (c) outputs, (d) activities, and (e) inputs.

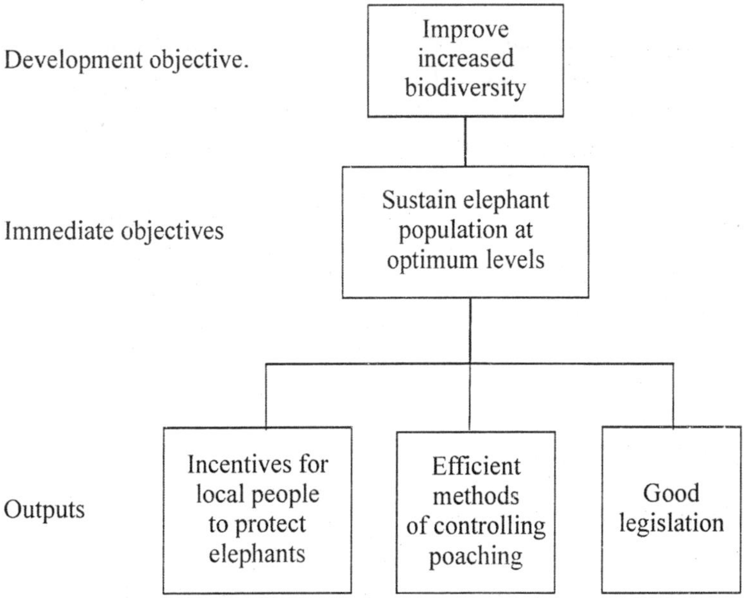

Fig. 7.3: Strategy Analysis

Development objectives : It is also called as broad objective or goal. It describes the development benefits which the respective target groups can expect to gain from the project or programme. It contains hints on the kind of benefits which are expected to accrue to the target groups and by what type of capabilities they should be enabled to keep up or improve their conditions in changing economic, social and institutional environments. Therefore, it is a long term goal that a project/ programme aims to achieve in synergy with other development interventions (*The emphasis here is evaluation of impact that is, effect).*

Immediate objectives : Also called as direct objectives or project purpose. The purpose of a programme or a project describes the changes in behaviour, structures or capacity of the target groups which directly result from the utilization of the deliverable outputs or results of the programme or project is expected to yield. Thus, it is the specific change in behaviour or condition, to be achieved at the end of project/programme *(The emphasis here is evaluation of effect that is, effect-cause).*

Outputs : Also called as results, which describe the goods and services, the direct deliverables which are contributed from the side of a project or programme. The outputs or results must express the nature, scope and intensity of support or the solution being sought. Thus, it is tangible result of project's activities which provide the opportunity for change or progress *(The emphasis here is monitoring of results that is, effect-cause).*

Activities : Measures or tasks carried out by the project/ programme in order to achieve and obtain the outputs/results. Therefore, it is specific substantive tasks performed by the project staff or associates *(The emphasis here is monitoring that is, effect-cause).*

Inputs : Also called as resources, which are identified as necessary means for performing activities. All project resources provided by the participating parties *(The emphasis here is monitoring that is, cause).*

(ii) Objectively Verifiable Indicators

For each cell of the narrative summary, objectively verifiable indicators need to be developed. They should be SMART. It means that the indicators should be specific, measurable, attainable, relevant and timely.

Specific : Key indicators need to be specific and be related to the conditions the activity seeks to change.

Measurable : An indicator must be able to be measured in either quantitative or qualitative terms. Quantifiable indicators are preferred because they are precise, can be aggregated and allow further statistical analysis of the data. However, development process indicators may be difficult to quantify and qualitative indicators should be used.

Attainable : The indicator must be attainable at reasonable cost using an appropriate method. It should be feasible in terms of finances, equipment, skills and time.

Relevant : Indicators should be relevant to the management information needs of the people who will use the data. Also, an indicator should reflect what we are trying to measure in an accurate way as well as, capable of picking up changes over a time period.

Timely : An indicator should be able to provide information in a timely manner.

(iii) Means of Verification

Once the indicators are developed, the source of information, method of data collection, method of data analysis, periodicity of data collection and resources need to be established for each indicator.

(iv) Assumptions

For each item of the narrative description, assumptions have to be specified. The purpose of specifying assumptions is to : (a) assess the potential risks to the project concept right from initial stages of project planning, (b) to support the monitoring of risks during implementation of the project, and (c) to provide firm basis for necessary adjustments within the project whenever it is required. Therefore, these are mainly external factors/conditions having bearing on implementation of a project over which project managers have no direct control. To identify assumptions, one has to go through the hierarchy of project elements from bottom to top, asking the question; what else, according to our knowledge and experience, needs to be in place (factors outside the project's control) before we reach the next level?

To control the accuracy of an assumption, one has to ask three questions.

Q_1 Is the assumption really important?
If yes, go to the next question
If no, don't include in the planning matrix

Q_2 How likely will it occur?
If it is almost certain—don't include
If it is quite likely—include, monitor and possibly influence
If it is not likely—go to the next question

Q_3 Can the project strategy be (re) designed to make this assumption superfluous?
If no, killer assumption! Stop!
If yes, redesign the project and plan activities to influence assumptions

LFA/Project Planning Matrix

Narrative Summary	*Objectively Verifiable Indicators*	*Means of Verification*	*Important Assumptions*
(a) Development Objective	What are the quantitative ways of measuring or qualitative ways of judging, whether these broad objectives are being achieved?	• Sources of information • Methods of data collection • Methods of data analysis • Periodicity of data collection • Resources needed	• What external factors are necessary for sustaining objectives in the long run?
(b) Immediate Objectives	What are the quantitative measures or qualitative evidence by which achievement and distribution of impacts and benefits can be judged?	• Sources of information • Methods of data collection • Methods of data analysis • Periodicity of data collection • Resources needed	(immediate objectives to broad objective) • What condition external to the project is necessary if achievements of the project's purpose are to contribute to reaching the project goal or broad objective?
(c) Outputs	What kind of quantity of outputs and by when will they be produced? (quantity, quality, time)	• Sources of information • Methods of data collection • Methods of data analysis • Periodicity of data collection • Resources needed	(outputs to immediate objectives) • What are the factors not within the control of the project which, if not present, are liable to restrict progress from outputs to achievement of immediate objectives?

(*Contd.*)

(*Contd.*)

Narrative Summary	*Objectively Verifiable Indicators*	*Means of Verification*	*Important Assumptions*
(d) Activities	What kind of specific tasks performed by the project staff?	• Sources of information • Methods of data collection • Methods of data analysis • Periodicity of data collection • Resources needed	(activities to outputs) • What external factors must be realized to obtain planned outputs on schedule? • What kind of decisions or actions outside the control of the project is necessary for inception of the project?
(e) Inputs	What types and levels of resources provided by the participating parties?	• Sources of information • Methods of data collection • Methods of data analysis • Periodicity of data collection • Resources needed	• Main external factors/conditions bearing on the implementation of a project over which project managers have no direct control

The Log Frame and Monitoring and Evaluation

The log frame helps to establish the basis for monitoring and evaluating the activity. The link between the log frame and monitoring, review and evaluation is illustrated below :

Log Frame Hierarchy	*Type of Monitoring and Evaluation*
Broad objective (Impact)	Ex-post evaluation
Immediate objectives (Project purpose/outcome)	Evaluation at completion
Outputs	Monitoring and review
Activities	Monitoring
Inputs	Monitoring

D. Implementation

The operational phase of a project commences when implementing activities begin in order to achieve the expected outputs/results. In most cases, this may be one or two years after the project concept had been established at the end of the design phase (PPM). In the mean time, log frame conditions may have changed, so that a verification of the PPM must take place during the operational planning.

Implementation should have a plan of operations and it should be established by the project team and will be documented as: (a) work plans/work schedules, (b) project budget/resource plans, (c) personal plans, (d) material and equipment plan/procurement plan/staff training plans.

Steps in construction of a work plan-cum-project budget

Step 1: Develop a task (work plan skeleton) and transfer the activities from PPM to first column.

Step 2: For each activity/sub-activity: (a) specify anticipated milestones, (b) assign responsibilities for the implementation, (c) determine the timing of the activity (beginning and ending), (d) specify staff requirement, (e) specify the quantity of material/equipment needed, (f) specify costs, and (g) specify important assumptions.

Step 3: Final check, with focus on: (a) consistency of costs with overall allocation, (b) workload of responsible personnel, and (c) consistency of the timing. It is advisable to transfer the work plan into a bar chart with time axes.

Advantages of LFA

- It brings together in one place a statement of all key components of a project (helpful when there is a staff change);
- It presents them in a systematic, concise and coherent way, thus, clarifying and imposing the logic of how the project is expected to work;
- It separates out the various levels in the hierarchy of objectives, helping to ensure that inputs and outputs are not confused with each other;
- It clarifies the relationships which underlie judgments about likely efficiency and effectiveness of projects;
- It identifies the main factors related to the success of the project;
- It provides basis for monitoring and evaluation by identifying indicators of success and a means of quantification or assessment ; and
- It encourages multidisciplinary approach to project preparation and supervision.

Limitations of LFA

- The LFA is often developed and used rigidly. This can stifle innovative thinking and adaptive management; and
- The LFA do not readily enable monitoring unintended consequences.

Conclusions

The LFA provides a powerful set of tools for designing projects and project evaluations. However, like all tools, LFAs are not the complete answer to effective project designing. The LFA is best used towards the end of the project design cycle after information has been collected and analyzed, needs are assessed, views of stakeholders sought and the external environment of the project is understood.

8

Farmers Field School : Implementation Guidelines for Participatory Learning

N. Nagaraja and *C.M. Savitha*

Introduction

Sustainable farming requires an enhanced capacity of farmers to adopt in the face of unexpected changes and emerging uncertainties, rather than mere continuation of a practice forever. Helping farmers to enhance their abilities to understand and analyze their situation enabling good decision making is very important. Farmers Field School (FFS) is one of the established participatory methods of effective learning. Farmers Field School was considered as an effective and comprehensive non-formal educational method to teach and technically empower adult farmers and farm women. Farmers as adult learners, have unique advantages as well as some particular handicaps. They have some specific needs to be met for which they are willing to make purposeful efforts. They can evaluate better and integrate their learning with past experience more easily. Once they are interested in something, convinced of its usefulness and gain self-confidence, they become as good learners as anybody else. Thus, Farmers Field School as a non-formal education and a learner centred educational process was designed to make them to learn and technically empower by creating a strong desire for learning, to solve their problems by setting clear goals, stimulating their thinking to solve the identified problem, creating situation to put forth efforts to learn by themselves, to solve the problem by testing the identified solution and thus, made to derive satisfaction in their learning process by avoiding situations which frustrate them.

FFS was designed initially as a way to introduce Integrated Pest Management (IPM) practices to irrigated rice farmers in Asia. The FFS approach was first developed by FAO in Indonesia in 1989 for training rice farmers on integrated pest management. It was developed as a way for small-scale rice farmers to investigate and learn for themselves the skills required for, and the benefits of adopting integrated pest management practices. From 1991 to 1994 with support from FAO inter-country IPM programme, rice IPM FFS expanded to Bangladesh, Cambodia, China, India, Philippines, Sri Lanka and Vietnam. In Africa the FFS concept was introduced (1994) in Ghana and currently being implemented in many countries like Burkina Faso, Mali, Zambia, Zimbabwe, Kenya, Uganda, Tanzania, DRC, Gambia, Mozambique, Ethiopia, Malawi, Sudan, Nigeria, and Rwanda.

From 1995 the FFS programme began to broaden the scope beyond IPM and started to cover other types of production and incorporated Socio-ecological conditions. FFS approach relies on participatory training methods to convey knowledge to the participants and equip them technically.

What is Farmers Field School?

Farmers Field School (FFS) is a non-formal learner centred educational process. It seeks to empower people to solve their field problems actively by fostering participation, interaction, dialogue, joint-decision making, self-confidence and self-determination. As the name suggests Farmers Field School is a school for the farmers outside the class rooms and on the farms. It operates with the principles of non-formal education and most of the sessions and contents based on the adult learning principles.

FFS is not about technology; but, it is about people development. It is a community based, practical oriented field study programme involving a group of farmers facilitated by extension staff or by other farmers. It is the forum for farmers to learn, test and adopt practices using practical hands on methods of discovery learning with emphasis on observation, discussion, analysis and collective decision-making.

Why Farmers Field School?

In FFS, field where the specific crop is cultivated is the learning site. Therefore, it is possible for the farmers to get first hand experience. It centres on the identification of on-farm problems through participatory approach. Participants of FFS work in small sub-groups; collect data, make analysis of the data and take decisions based on the results obtained, and hence, they get technically empowered. FFS emerges as a learning activity which promotes healthy and quality decision-making. Farmers conduct their own field studies with comparisons or treatments, and in the process, learn how to be in groups and sharpen their experimental skills.

It involves learning throughout the crop season and, therefore, enables farmers to get deeper insights regarding their crop ecosystem through individual evaluation. It provides ample opportunities for Experiential Learning, and helps in building self-confidence and promotes farmer-to-farmer communication resulting in speedy diffusion of technologies.

Key Features/Characteristics of Farmers Field School

(i) *Learning by Doing :* Farmers "learning-by-doing" i.e., they carry out for themselves the various activities related to the particular farming practice they want to study and learn about. It is based on the principle that "If I hear it I forget it", "If I see it I remember it" and "If I discover it I can own it for life". This could be related to any area like annual crops, livestock, soil fertility, water management, etc.

(ii) *The Field is the Learning Place (School Without Walls). Farmers do Everything :* Farmers learn directly from what they see, collect and experience and not from a text book, picture or other extension materials. Farmers produce their own learning materials. The advantage of these home made materials is that they are consistent with local conditions, less expensive and owned by the farmers. Members of each working group assume responsibility for presenting their findings in turn to their fellow farmers.

(iii) *Facilitation, NOT Teaching by Extension Workers :* In FFS extension workers are called facilitators because their role is to guide the learning process "make it easy" and not to teach. The facilitator contributes to the discussions aiming to reach consensus on what actions to be taken.

(iv) *Season Long Comparative Experiments :* Training is based on comparative studies and field studies conducted by farmers. Training is related to the seasonal cycle of the practice being investigated, e.g., seed to seed, egg to egg, hydrological cycle, etc.

(v) *Regular Group Meetings :* Farmers meet at agreed regular intervals, and it could be weekly, fortnightly or monthly depending on enterprise or activity.

(vi) *Farmers Decide the Curriculum :* Farmers decide what is relevant and need to be addressed in the FFS, where the facilitator only guides the farmers in their learning process by creating opportunities for experiences.

(vii) *It is a Systematic Training Process :* FFS follow a systematic training process. The key steps are observation, group discussion, analysis, conclusion and action plan. The training process uses agro-ecosystem analysis (AESA) and is developed gradually during regular and frequent meetings. Experience shows that better results are achieved with weekly meetings than with less frequent FFS sessions.

(viii) *Every FFS is Unique :* Training activities have to be based on the gaps in farmer's knowledge taking into consideration their level of understanding. Every group is different with different needs and realities. Participants develop their own learning content and hence each FFS is unique.

What are the Subjects for a FFS?

The agriculture related problems faced by the farmers make subject for FFS. It could be low yield, nutrient management, pest management, water management, etc. Field operations in FFS plot takes place in parallel with the operations carried out in the farmers' fields. For example, paddy transplanting in FFS takes place at the same time as farmers are transplanting their own crops. Therefore, lessons learnt in FFS can be applied directly in the respective fields.

Actors in FFS and their Desirable Qualities

Mainly three categories of actors are involved in Farmers Field Schools and they are: (1) Participants, (2) Collaborator and (3) Facilitator.

1. Participants

These are the farmers selected by villagers, based on certain criteria and willing to be the learners, experimenters in the Farmers Field School. Enough representation for woman participants (about 33%) is always encouraging. Must participate in all the sessions (once in 10 days) throughout the crop season. Must involve actively during field school activities and adopt the efficient practices identified in the FFS in his/her field in subsequent season/year and continue to adopt in later years also. Must share the knowledge/skills learnt in FFS with other farmers, and so must be able to read and write.

2. Collaborator

Collaborator is a farmer/farmwoman who gives land and other facilities/material for conducting field studies. Collaborator should be cooperative, preferably non-political, volunteer to spare his/her land and bear the input costs, allow conducting FFS in his/her field, be available during FFS sessions or be one of the FFS participants.

3. Facilitator

Facilitator is a technically competent person to lead members through the hands-on exercises. A good facilitator should possess social, communication and organizational skills to handle FFS effectively. The facilitator will participate in discussion sessions as a contributor rather than a leader in arriving at an agreed consensus. Local extension personnel will act as facilitator. He/she will encourage farmers to identify their field problems, search for solutions, and test them in the field to solve such problems. In this regard, they develop plan of action and implement as per plan. Facilitator makes the farmers to carry out the activities. Once, the farmers are made to know what they have to do and what they have to observe in the field, the facilitator takes a back seat. Facilitators are expected to have following qualities:

(i) Technical Competence

- Possesses complete knowledge of the technology selected for FFS;
- Skilful in carrying out all operations related to the selected technology;
- Skilful in diagnosing the problems and solving them;
- Skilful in establishing cause and effect relationship for each activity/interventions selected in FFS;
- Skilful in conducting field trials; and
- Skilful in drawing valid inferences/conclusions on the studies/observations.

(ii) Social Skills

- Should be a good leader;
- Capable of working with groups by understanding group dynamics (both internal and external dynamics);
- Accept the suggestions given by the participants;
- Ensure active participation of all the members in the group;
- Promote group decision-making by consensus;
- Resolve conflicts likely to arise;
- Enters into the field earlier than others.

(iii) Communication Skills

- Willing listener;
- Keen observant;
- Ensures effective flow of information;
- Stimulates thinking among participants;
- Ensures equal opportunities to all the members to express their views/opinion; and
- Promote good discussions/interactions among members.

(iv) Organizational Skills

- Skilful in setting the objectives;
- Skilful in developing strategy to achieve objectives;
- Skilful in planning, execution and management;
- Skilful in motivating the participants; and
- Skilful in getting cooperation of all the local organizations/ institutions.

Limitations of Farmers Field School

- Ensuring attendance of all the participants for all the sessions will be difficult;
- It can create a negative impact on villagers if FFS fails to achieve expected results;
- It is a time consuming activity;
- It demands lot of preparations on the part of the facilitators; and
- It is a costly extension education method.

Steps Involved in Organizing a FFS

The FFS to be effective, should be organized systematically by adopting nine steps in a sequence. The details are described below :

Step 1. Enlisting Cooperation of the Development Departments/ Organizations

As a first step, it is essential to get the basic information related to crops, specific problems of the farmers where the FFS has to be organized. He/she has to contact extension staff of the Development Departments/organizations and develop rapport with the staff, enlist their cooperation in carrying out FFS process.

Step 2. Sensitizing the Villagers about FFS

Concept and importance of FFS need to be explained to villagers in a discussion meeting. FFS alumni can be used in such discussion session (if available). The importance of FFS in solving their problems by themselves needs to be emphasized (Photo 8.1).

Photo 8.1: Sensitization of farmers by project workers

Step 3. Selection of FFS Participants and Collaborator

Villagers should be given criteria for selection of about 30 FFS participants and one or two collaborators. The facilitator (extension personnel) should clearly explain each criterion to the villagers. Then, villagers should select the candidates (participants) and collaborator by considering the set criteria.

Step 4. Development of Norms for FFS

After the selection of FFS candidates and collaborator, the facilitator should encourage FFS candidates, collaborator and villagers to evolve norms/rules and regulations on regularity and punctuality, behaviour, resolving conflicts likely to arise, penalty for indiscipline/ misbehaviour, sharing the responsibility, etc. The norms evolved should be finalized by facilitator after discussion of each norm with FFS participants, collaborators and the villagers.

Step 5. Approval of FFS Participants, Collaborator and Developed Norms in Grama Sabha

List of selected FFS candidates, collaborator, and norms developed have to be presented to the Grama Sabha (village assembly) and after explaining the importance of FFS in the context of agricultural development, the approval of Grama Sabha for selected list of candidates, collaborator and norms has to be sought. This will be helpful in smooth implementation of FFS.

Step 6. Development of MoU

Before actual implementation of FFS, it is suggested that all the partners likely to be involved in FFS enter into a MoU indicating, each partner's commitments, roles, responsibility, punctuality, discipline, etc. This should be in accordance with the already approved norms for FFS in the Grama Sabha.

Step 7. Planning for FFS

Planning for FFS is crucial to its success. A number of activities have to be carried out to complete the planning process. The activities include (a) problem analysis, (b) topic selection, (c) development of treatment plan, (d) designing and layout of FFS field, (e) listing of material required, (f) scheduling of FFS activities, (g) preparation of calendar of operations, and (h) preparation of sessions guide. A brief description on these activities is given in the following paragraphs.

7.1 Problem Analysis

The problem analysis is the first step in planning process. List the problems faced by the farmers in that area, may be related to water management, nutrient management, pest management and other production related problems. Prioritize the problems and then select the technology for studying in FFS by employing SWOT analysis method (Photo 8.2).

Photo 8.2: Prioritization of problems by the farmers

7.2 Topic Selection for FFS

The technology selected for FFS will be based on the crops grown and problems encountered. Generally technology can be of three types, viz.

1. Integrated Crop Management (ICM): It emphasizes on adoption of integrated approaches in management of water, nutrients, pests and weeds, to increase yields, to reduce environmental ill effects and cost of cultivation.
2. A special topic: which requires investigation to solve the existing problem. For example, paddy cultivation with different intensities of irrigation, which has to be tested by conducting trials of paddy cultivation as beneficial to them in saving water and supplying water to tail end farmers.
3. Current practices followed by farmers which will be used to compare with trials.

Technology Selection Matrix

Problem Faced by the Farmers	*Current Practices being Followed with respect to the Problems*	*Potential/ Opportunities for Improvement*	*Reasons for not Implementing the Potential Practices Earlier*	*Selection of Topic for FFS Study*
Ex. Tail-end farmers are getting inadequate water during later stages of paddy crop	• Farmers in head reach grow transplanted paddy with 3-4" water submergence in the field throughout the crop period	• Changing the irrigation practice, i.e., instead of submergence wetting the paddy field.	• Lack of awareness about advanced irrigation method (wetting method)	• Light irrigated paddy cultivation or Aerobic method of paddy cultivation
	• Growing longer duration varieties	• Cultivation of short duration paddy varieties	• Lack of availability of short duration varieties	• Selection of short duration varieties

The facilitator should see that the need for conducting studies (especially for the special topic) on the above technologies come from participants. Under no circumstances the facilitator should indicate or force the participants to go for these technologies. Efforts should be made in such a way that the need arises from the participants themselves. For this purpose, technology selection matrix as indicated above may be adopted.

7.3 Development of Treatment Plan

After selection of technology, treatments have to be decided. Ideas of the participants have to be pooled and analyzed for finalizing the treatments. The idea matrix indicate the possible treatments for testing. For example, some farmers say wider spacing is better or some may say there will be reduction in the yield due to wider space. Likewise, pool all the ideas and analyze them.

Idea Matrix

Idea to be Tested	*Source of Idea (Name of the person who gave the Idea)*	*What has to be done with the Idea as expressed by the proposer?*
Ex. Light irrigation to paddy crop will yield more	Farmer X	Need to be tested
Light irrigation to paddy crop will increase weeds	Farmer Y	Need to be tested
Light irrigation to paddy crop increases the pest incidence	Farmer Z	Need to be tested
Light irrigation to paddy crop reduces the tillering capacity	Farmer A	Need to be tested

Based on the above ideas, establish practice cause-effect relationship and also state the hypotheses for the testing, using the matrix given below. After developing the hypotheses for testing, decide the treatments and replications to be tried and accordingly, finalize the design and layout for the study. Number of trials depends on the hypotheses set for the study. An example of trials stated in the practice cause-effect relationship matrix is given below.

Practice cause effect relationship matrix

Practice	*How does it help to improve the present situation ?*	*Possibility of operationalising the practice*
Ex.Light irrigated paddy cultivation in place of continuous submergence	Less water requirement for light irrigated paddy helps to save water and thus saved water could be used for other area	It is possible to practice light irrigated paddy cultivation

Based on the above matrix following hypothesis could be developed :

Hypothesis

(i) Paddy cultivation through light irrigation method in tank command area in place of continuous submergence will help to save considerable quantity of irrigation water.

(ii) Changing paddy variety from longer duration to shorter duration will help to get expected yields in the event of less water availability in the tank.

(iii) The yields/unit area will be more if ICM practices in paddy are adopted.

Examples of Trials based on the above Hypotheses

(i) Paddy cultivation with different irrigation intensities.

(ii) Cultivation of paddy varieties of different duration.

Treatments for First Trial

T1 Submerged paddy cultivation.

T2 Punaji method of paddy cultivation.

T3 Light irrigated/Aerobic paddy cultivation.

Treatments for Second Trial

T1 Tella Hamsa Variety.

T2 Rasi Variety.

T3 KRH-2 Hybrid.

7.4 Design and Layout of FFS Plot

Select about 0.8 ha of land in the tank command area, belonging to one or more farmers. They will be the collaborators in the FFS. Out of

Photo 8.3: A view of treatments laid out

0.8 ha of land, 0.30 ha should be allotted for Integrated Crop Management practices, 0.20 ha for conducting field experiments/trials based on the problems observed in the locality with a check plot/ control plot as well as a general plot, and in the remaining 0.30 ha, the farmers practices could be continued. The design and layout may be as given below :

Layout of FFS Field

<table>
<tr><td rowspan="6">Integrated Crop Management Practices</td><td colspan="8">General Plot</td><td rowspan="6">Farmers Practices</td></tr>
<tr><td colspan="8">Trial Plot with a Maximum of Three Treatments Example</td></tr>
<tr><td colspan="4">Water Management Practices</td><td colspan="4">Varietal Trial</td></tr>
<tr><td>R_I</td><td>T_1</td><td>T_2</td><td>T_3</td><td>R_I</td><td>T_1</td><td>T_2</td><td>T_3</td></tr>
<tr><td>R_{II}</td><td>T_2</td><td>T_3</td><td>T_1</td><td>R_{II}</td><td>T_2</td><td>T_3</td><td>T_1</td></tr>
<tr><td>R_{III}</td><td>T_3</td><td>T_1</td><td>T_2</td><td>R_{III}</td><td>T_3</td><td>T_1</td><td>T_2</td></tr>
<tr><td>030 ha-----</td><td colspan="8">-----------------0.20 ha-----------------</td><td>-----0.30 ha-----</td></tr>
</table>

7.5 Material Required for the Trial

Based on the topic selected for the trial, list of materials, quantity and time of requirement may be detailed as mentioned below :

Material required for FFS

Sl. No.	Material	Quantity Required	Time of Requirement	Source of Supply
Ex: 1.	Seed			
2.	Manures and Fertilizers			
3.	Plant protection chemicals			
4.	Micro nutrients			

7.6 Decision on Recording Observations

During every visit (session), observation on the trials to be carried out and method of recording the observations has to be decided, and for this purpose observation matrix has to be developed as outlined below :

Observation Matrix

What to Observe?	How to Observe?	When to Observe (Stage of the Crop)?	Materials Required for Observation
Ex. 1. Growlh parameters			
• Plant height	Scale	30,45,60,75,105	Metric scale
• No.of tillers	Counting	30,45,60,75	–
• No.of leaves	Counting	30,45,60,75	–
Ex. 2. Yield parameters			
• No.of panicles hill	Counting	Panicle initiation	–
• No.of grains/panicle	Counting	Grain setting	–
• 100 seeds weight	Weighing	After threshing	Balance
• Chaffy grains	Weighing	After threshing	Balance

7.7 Scheduling the Visits of FFS

The facilitator should encourage the FFS participants and collaborators to decide about frequency of visits to FFS site based on the technology considered. The scheduling of visits should start from first stage of land preparation till harvest of the crop. The frequency could be once in a week or once in 10-15 days. In any case, they should agree and decide on it. Also, for each visit, the time of assembling at the field need to be specified. It is advisable to visit field from 8.00 am to 1.00 pm to get some valuable data for agro-ecosystem analysis (AESA).

7.8 Preparation of Calendar of Operations

After scheduling the frequency of visits, next stage would be detailing the activities to be carried out during each visit/session. Keeping the crop and treatments/technology to be implemented, the facilitator should encourage FFS participants and collaborator to detail the activities for each of the scheduled visit/session. It should be discussed and finalized. The relevant format is given below :

Calendar of Operations

Session	*Week /Date*	*Activity to be carried out*
1		
2		

7.9 Session Guide

Is is a blue print to be prepared after finalization of calendar of operations. For each session a session, guide has to be prepared and it should have details on title of the session, introduction, time required, material required, methodology, discussion questions and conclusion. An example of a Session Guide is given below in the next page.

Step 8. Execution of FFS

As per the calendar of operations, the activities should be started and the predetermined frequency of conducting sessions has to be

An Example of a Session Guide

Session Title	: Conducting Ballot Box test
Introduction	: There is a need to test the existing knowledge of farmers on a particular subject. Normally this is conducted prior to initiation of any activity.
Objective	: To know the extent of knowledge possessed by the farmers on the technology under consideration for FFS.
Time required	: 25 minutes.
Materials required	: Card board, drawing sheets, gum, pen, stick, scissors, specimens and thread

Methodology

- Preparation of ballot box using card board and drawing sheets.
- Preparation of questions with multiple choices for each question.
- Write each question on separate ballot box (at the rate of one box for each question) and the number of pockets in a ballot box should be according to the number of multiple choices given for that particular question.
- Place the ballot box in a circular manner at 5 feet interval
- Allot serial numbers to the farmers and issue slips indicating their serial number. Each farmer should be given serial number slips equal to number of questions included in the test.
- Request farmers to stand in front of the ballot box.
- Give instructions to farmers to insert his/her serial number slip in the relevant pocket of the ballot box. Each ballot box will have pockets equal to multiple choices given. The farmer should choose the pocket based on his answer. For example for a question, there are three multiple choices as *Yes, No, Do not know.* If his answer is Yes, he/she should insert the serial No. slip into *Yes* pocket.
- Allot equal time for each question and ask the farmer to move to next ballot box after the specified time.
- Count the correct answers and work out the average knowledge possessed by the group on the subject.

Discussion questions :

- Discuss about the conduct of test.
- Needed changes, if any.
- Its usefulness, etc.

Conclusion :

Based on the test results, the facilitator can plan the future activities.

carried out. During each visit or session, the FFS candidates have to be divided into sub-groups of 5-6 farmers. Each sub-group will have a leader and will study crop ecosystem and trials during every session Each session will be of about 5 hours. This time has to be used very productively. The suggested break-up of the stipulated time is as follows.

About 15 minutes recap and briefing to give continuity to the previous week's work. Next 30 minutes should be used for agenda setting for the present day visit. Here, details of activities to be carried out and outputs to be achieved will be decided by the group. About 2 hours will have to be allocated for field activity. Field activity includes carrying out relevant field activities, agro-ecosystem analysis, collection of plants samples, insects, specimens, etc., recording observations as decided and indicated in observation matrix. In agro-ecosystem analysis, ecological interactions between the environment, plants, herbivores (diseases, insects and rats) and natural enemies of herbivores take place. The herbivores are balanced by natural enemies. The inter-dependencies of soil, water and nutrients; need to be understood well, to know when to irrigate, how much to irrigate, when to take pest control measures to prevent economic losses and maximize profits. Next 30 minutes should be allocated for processing of agro-ecosystem analysis. Afterwards, the observations and assessment in trial plots have to be depicted diagrammatically by the participants themselves on a chart paper. The diagram should depict the field condition, growth stage of plant, weather conditions during past one week, moisture/water level, weeds, pest population, natural enemies population, diseases, past experiences, investment costs, etc. After agro-ecosystem analysis, the sub-groups discuss on the observations made in the trial plots and prepare a note on their observations and get ready for presentation in the general session. A specimen copy of the outline of AESA outcome chart is given below. About 30 minutes to be devoted for conducting group exercises in the form of games to break the monotony. This should be followed by presentation of each sub-group output by respective leader on the observations made followed by discussion. This will last for 30 minutes. Thereafter, 30 minutes should be given for finalizing activities for the next session and listing the material/equipment required. At the end, a review of day's activities has to be undertaken by the facilitator in 15 minutes time.

AESA Chart Presentation Model

AESA Agro Ecosystem Analysis

Date: Weather conditions Synny/cloudy/rainy (Pictorial)	Crop : Variety :	Group No. : Crop stage : 60 Days
Harmful insects & Diseases (Pictorial)	Picture of the Crop Plant indicating Plant no. of branches, flowers, fruits & yield parameters to be drawn in this place	Beneficial Insects/ Predators & Weeds (Pictorial)

Points Discussed	Decisions taken
e.g. : • Soil moisture level is very low.	• Immediate irrigation required.

8.1 Recording Observations in the Trial Plot

After carrying out agro-ecosystem analysis, let each group record the observations as decided earlier. The information is gathered by sampling which involves closely observing and monitoring small part

Photo 8.4: Recording of observations

of a large field (trial plot). The aspects of crop ecosystem as detailed in observation matrix are to be observed. The sampling results are used to arrive at decisions based on crop health status, natural enemy ability, probable yield, etc.

8.2 Analysis

Analysis has to be carried out at all stages of the crop as specified in the observation matrix. While carrying out analysis, simple

Photo 8.5: Participants compare the treatments

comparisons need to be undertaken for each treatment by drawing lines. It is called as overlapping test. An example of analysis of plant height is given below :

Analysis of Plant Height using Overlapping test												
Ex: Plant height in cm. At 30 days of planting												
			T_1	T_2	T_3							
		R_1	30	35	40							
		R_2	35	32	37							
		R_3	30	35	45							
	T_3	....										
	T_2	....										
	T_1	....										
	30	31 32 33 34 35	36	37	38	39	40	41	42	43	44	45
..................................*Plant height in cm*..................................												
From the above analysis, it could be inferred that there is no remarkable difference between T_1 and T_2 with respect to plant height, whereas T_3 is superior over T_1 and T_2												

8.3 Organizing Field Day

With a view to educating large number of potential farmers, a field day has to be organized when the results of the FFS activities are observable and significant. The FFS participants and collaborator should present the whole process to the participants in the field day, besides showing the field to the visitors and explaining the details of activities carried out.

Photo 8.6: A Field Day in Progress

8.4 Final Evaluation

At the end of the trial, the ideas tested have to be evaluated to draw meaningful conclusions. Table indicating the values to draw valid conclusions is given below :

Evaluation of Treatment Effects

Sl. No.	*Treatment*	*Parameters*				
		Growth		*Yield*		*Inference*
		Plant Height (cm)	*Tillers (No.)*	*Panicles (No.)*	*Grain Yield (q/ha)*	
1.	T_1	95	45	40	35	Though grows up to 95 cms. gives higher yield/unit area
2.	T_2	100	40	30	25	Grows taller, hence it gives more straw but less grain yield
3.	T_3					

Conclusion

Based on the inferences, let the FFS participants deliberate and decide about the best treatment to be adopted on a regular practice in the coming years.

Step 9. Withdrawal

After the harvest of crop, the results of the activity have to be analyzed during the last session of the FFS activity. After presentation of results by FFS candidates and collaborator, they should be encouraged to follow the same technology in their own fields and also to influence

Photo 8.7: Summing up and setting the future road-map

their neighbours; and friends to adopt the technology. The required help and other support need to be finalized and extended as per the need. The outcome of FFS needs to be made known to all potential farmers. For this it may be advisable to write the results of FFS/ outcome of FFS on the wall at a prominent place in the village so that it can serve as reminder.

9

Participatory Technology Development : Experience of Karnataka Community Based Tank Management Project

N. Nagaraja and *N. Manjula*

Participatory Technology Development (PTD) : Meaning and Importance

Participatory Technology Development (PTD) is essentially a process of purposeful and creative interaction between rural people and outside facilitators. Through this interaction, the partners try to increase their understanding of the main traits and dynamics of the local farming systems, to define priority problems and opportunities, and to experiment with a selection of 'best bet' options for improvement. The options are based on ideas and experience derived from both indigenous knowledge (both local and from farmers elsewhere) and formal science. This process of technology development is geared not only towards finding solutions to current problems, but also towards developing sustainable agricultural practices which conserve and enhance the natural resources so that they can still be used by future generations. Most important of all, PTD should strengthen the capacity of farmers and rural communities to analyze ongoing processes and to develop relevant, feasible and useful innovations.

The process of technology development is closely linked with a process of social change. Encouraging local innovation through self-organized planning, implementation and evaluation of systematic experiments fosters self-respect and self-confidence in the rural communities involved. It also fosters a process of cultural awareness and change, as the planning and assessment obliges the participants to take account of their situation and the responsibilities of different people in the community (e.g., the different needs of women and men and the different barriers they face in trying to change their situation). Moreover, each community is made up of a variety of individuals with a diversity of interest and lifestyles, as becomes apparent especially when criteria for 'improved' technologies are debated. Yet another complication may be that natural resources are being used not just by one sedentary community but also by temporary immigrants and new immigrants into the area. Therefore, the process of local technology development in agriculture, horticulture and sericulture, etc., will also often involve negotiations and the development of new institutions to accommodate change in the use of natural resources.

Characteristics of PTD

The participatory technology development is a complementary process to the research conducted by the scientists. The PTD aims at linking the power and capacities of science to the priorities and capacities of farming communities to develop productive and sustainable farming systems. The scientists, extension workers and the farmers are the main actors in the participatory technology development. The scientists and extension workers act as external agency in the technology development process. The scientists and extension workers as an external agency, have to perform different functions at different stages. The success depends on the efficiency of the external agency in performing their functions.

Partners involved in PTD

Farmers, field extension staff of Development Departments, NGOs, input agencies, and Agricultural Research Stations are involved in the process. Out of which, farmers and field extension staff are involved during the entire process, others participate as and when required.

When to have PTD?

PTD is a season long process and is conducted, when (a) appropriate technologies are not available/known for a particular problem in a specific location, and (b) the existing technology involves high external input usage leading to high production cost and causing damage to the environment

Where to conduct PTD?

.PTD is done on farmers' fields, who are willing to take up trials. For purposes of comparisons, control plots are also maintained, which are similar to the plots in all the aspects including plot size.

Stages in PTD

There is a certain sequence in the way PTD is done. The PTD practitioners from outside the community introduce themselves to the communities where PTD process has to be started, build up a good relationship with local people, analyse the existing agricultural situation and form a basis for cooperation with networks of farmers to start the process of technology development. This includes, widening of understanding of all involved about the sociological, socio-economic, cultural and political dimensions of the current situation. There are six stages involved in PTD process, viz., problem identification, search for solution, designing experiments, conducting trials/experiments, assessment of research outcome and communication of results. For each stage, the approach to be followed, the functions of external agency and the expected outcome are described. The pattern of these stages is shown in the figure. However, the process under each stage need not necessarily follow the sequence in which it is described.

Problem Identification

Starter problem/core problem identification is the basic step in participatory technology development process. A careful analysis of both felt and unfelt needs or problems has to be carried out. Screening and discussing secondary data, community/farm walk, transect walk, problem census, community led surveys; focus group interviews, etc., are some of the techniques useful at this stage. The expected outcomes at this stage are: (a) an accepted agenda for PTD, (b) improvement in farmers' skills in analysing problems and looking for options, (c) increased social awareness and self-confidence, (d) PTD groups established, and (e) an improved organisational basis for conducting systematic local trials.

Example

Growing paddy (which is a high water consumptive crop) in the head reach of tank command area with continued submergence will result in non-availability of water to the crops in tail-end of the tank command area.

Search for Solution(s)

Document the indigenous knowledge/practices to the identified problem(s). Combining such knowledge/practices with the scientific knowledge and prioritizing the possible solutions are the important functions. Some of the approaches helpful to the scientists and extensionists to tap indigenous knowledge include making diagrams, case histories, critical incident techniques, preference ranking or matrix ranking. Farmer-expert workshops can also be organized. The outcomes at this stage are: (a) set of alternatives to solve the problems based on the SWOT analysis, and (b) improved skills of the farmers to analyse each alternatives based on the strengths, weakness, opportunities and threats analysis.

Example

To overcome the problem of non-availability of water to the farmers in the tail-ends of the tank command area, one of the solutions is changing the irrigation practice. i.e., instead of continued submergence, wetting the paddy field or cultivation by adopting light irrigation method or maintaining aerobic conditions or SRI method. The other solution is cultivation of short duration paddy varieties like KRH-2, Rasi, Mangala, Tellahamsa, etc.

Designing Experiments

It is important that the researcher design experiments that suit the farmers' purposes and strengthen their capacity to design experiments independently. It is important to determine the variables/parameters to be included in the experiment and their quantification techniques or means of verification. It is also essential to sequence the entire methodological aspects at this stage. Farmer-expert workshops, innovator workshops and study tours can be employed at this stage. The outcome of this stage are: (a) experimental designs that are reliable, evaluable and manageable by farmers, (b) improved skills of farmers to design trials and (c) monitoring and evaluation systems in place.

Example

To test the above two solutions in solving the problem, following treatments can be included in the experiment.

Treatments relating to Irrigation Management

T 1 = Submerged paddy cultivation (Farmers method);
T2 = Aerobic method; and
T3 = SRI method.

Treatments relating to Varieties

T1 = KRH-2;
T2 = Rasi;
T3 = Mangala

Number of replications = 3

The layout will be as given below :

Irrigation Trial				*Varietal Trial*			
R I	T 1	T 2	T 3	R I	T 1	T 2	T 3
R II	T 2	T 3	T 1	R II	T 2	T 3	T 1
R III	T 3	T 1	T 2	R III	T 3	T 1	T 2

Thus, 18 plots of convenient size have to be maintained.

Conducting Trials/Experiments

Calendar of operations has to be developed by involving all the partners and implementation has to be according to the calendar of operations. The step-wise implementation of the activities will strengthen the linkages among various agencies/organizations for

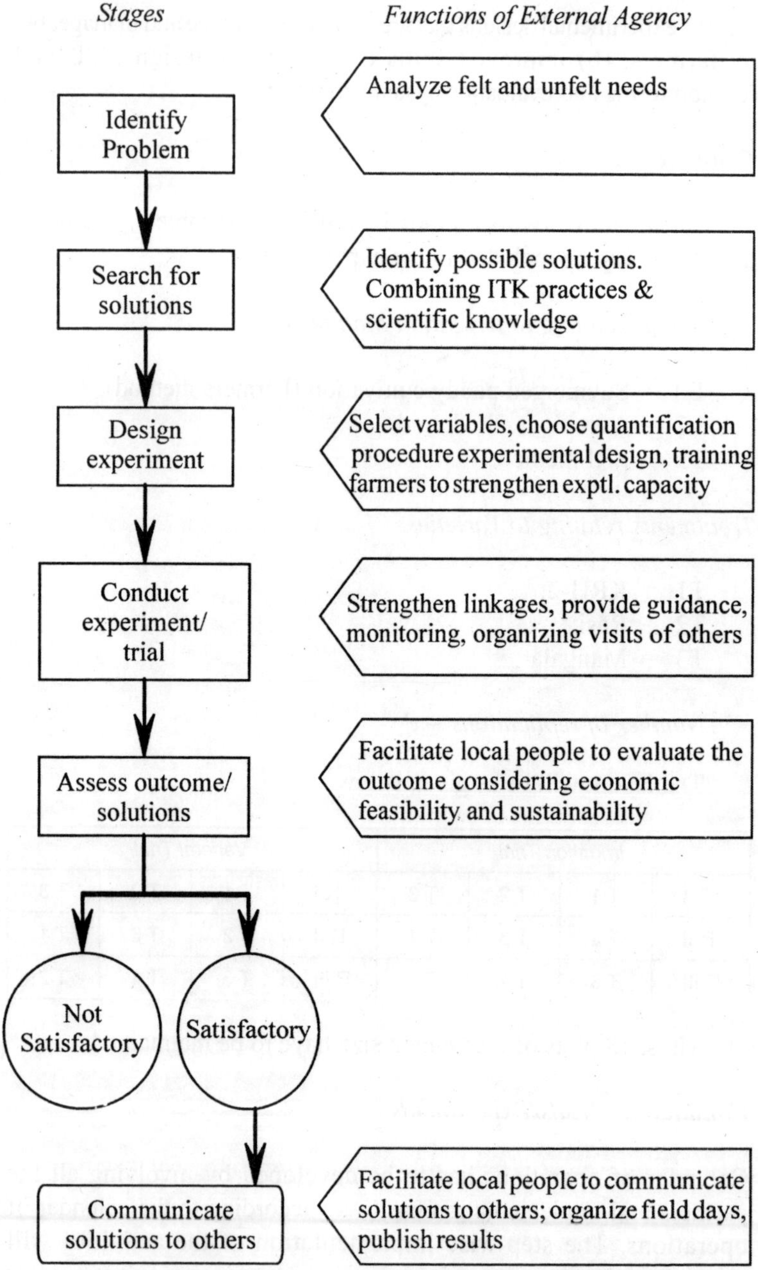

Fig. 9.1: Stages in Participatory Technology Development

timely and precise experimentation. Regular guidance to those involved in research has to be given. Especially, the farmers have to be trained in proper conduct of experiments. Group discussion meetings and method demonstrations involving concerned farmers at different stages of the crop growth have to be adopted. Proper monitoring is an important activity to be carried out by the scientists at this stage. The expected outcomes are: (a) a good number of experiments with technologies relevant to local situations are implemented and evaluated systematically, (b) development of PTD network, within and between villages, (c) institutional linkages are built, (d) the practical skills of involved groups of farmers to implement and evaluate experiments systematically are strengthened, and (e) growing active support of outside organisations and institutions.

Example: After initiating the experiments as per the layout and calendar of operations, recording observations on growth parameters, viz., plant height, number of tillers, number of leaves and yield parameters, viz., number of panicles/hill, number of grains/panicle, 100 seed weight, chaffy grains at appropriate stage of the crop, etc., have to be carried out jointly by the farmers, field extension workers and scientists. Appropriate tools for recording observations should be made available.

Assessment of Research Outcome

Assigning the farmers to objectively assess the outcome of research on economic feasibility such as relative advantage in terms of profitability and duration and sustainability factors viz. ecologically sound, economically viable, and socially just, humane and adaptable is necessary to prove the worth of the solution in the real world. If the outcome is not satisfactory, again the process should be repeated right from searching for solution stage. This should continue till the satisfactory solutions are identified. The important outcomes at this stage are: (a) identification of appropriate alternative to solve the existing problem, (b) future course of action on the outcome of the results, and (c) increased skills of the participants to assess the impact of different alternatives tried.

Example: After completion of the experiment, the observations recorded have to be analysed using simple and appropriate tests which the farmers can apply and draw valid conclusions. One such tests could be overlapping test where, simple bar diagrams are drawn for the mean scores.

Communication of Results

Sharing results with all potential farmers is the next essential activity. Farmer-to-farmer communication has been found-to-be more effective. The rate of diffusion will be faster if this technique is encouraged. Therefore, encouraging farmer-to-farmer communication is very essential. Some of the other techniques, which are useful, are organizing field days, group meetings, workshops at field level, publications, etc. The important outcomes at this stage are : (a) enhanced farmer-to-farmer diffusion of ideas and technologies, (b) an increasing number of villages involve themselves in process of organised technology development, making use of the experience of other communities, and (c) a farmer-managed system of inter-village training and communication.

Experience of PTD in Karnataka Community Based Tank Management Project

The University of Agricultural Sciences, Bangalore has entered into an agreement with the Jala Samvardhane Yojana Sangha—a nodal agency of the Government of Karnataka to render the consultancy services to increase productivity of 1500 tank command areas in the central and eastern dry zones of Karnataka, under the auspices of the World Bank funded Karnataka Community Based Tank Management Project. As part of the consultancy contract, the University has been carrying out the agricultural development activities including PTD. In three years' time (2003 to 2005), the University also introduced PTD process in 174 villages based on the problems identified in crops like paddy (77), sunflower (45), maize (15), groundnut (15), finger millet (3), and horticultural crops (19). The tank command area farmers, the staff representatives of community facilitation teams (NGOs), Department of Agriculture and the University were involved in the PTD process. The important topics covered were: (a) standardization of cultivation practices, weed management, simplification of sowing techniques, irrigation scheduling in aerobic rice cultivation, (b) integrated crop management approaches in sunflower, maize and finger millet, and (c) integrated pest and water management practices in groundnut and horticultural crops.

The significant outcome of PTD was standardization of cultivation practices for rice production under aerobic condition. The packages were first standardized in 10 hectares in 26 locations during 2003-04. Later, it spread to about 600 hectares in the project area by 2005-06. Similar results were also observed in maize, sunflower, groundnut and horticultural crops. The farmers as well as the project staff were enthusiastic in the process of experimentation at the village level. It has built the confidence of the field staff in technology identification/modification and dissemination. Hence, it proved to be an important approach to be tried by all those engaged in agricultural development process.

Conclusion

The participatory technology development process helps in identification of specific problem, search for alternative solutions to solve the problem, test the applicability of the alternative selected, and generate suitable techniques and practices to solve the existing problems. And this is done jointly by the farmers, extension functionaries and the scientists. Therefore, this process will cut short the time taken by the research system to identify the problem, conduct experiments and communicate results to the farmers through extension system. Also, this is more acceptable to the farmers and field extension workers as they are also the partners in the technology development process. However, participatory technology development is not a substitute for station based research or extension worker-managed on-farm trials. It is a complementary process, which involves linking the powers and capabilities of agricultural science to the priorities and capabilities of farming communities to develop suitable technologies to the problems observed.

REFERENCES

Nagaraja, N., 2003, *Raithara Kshethra Pathashale Anushthana Margadarshi* (Farmers field schools implementation guidelines), Gayathri Book Company, Bangalore.

———, 2004, *Areneeraavari (aerobic) Battada Besayada Anubhavagalu*, Gayathri Book Company, Bangalore.

10

Theory-based Evaluation and the Application of Grounded Theory in the Extension Context

G.M. Varadaraju and *M.J. Chandre Gowda*

Theory-based evaluation has similarities to the Log Frame Approach but allows a much more in-depth understanding of the working of a programme activity. In particular, it need not assume simple linear cause-and-effect relationships.

Theory based evaluation provides early feedback about what is or is not working, and why. It allows early correction of problems as soon as they emerge, and assists identification of unintended side-effects of the programme. It also helps in prioritizing which issues to investigate in greater depth, perhaps using more focused data collection or more sophisticated Monitoring and Evaluation techniques. In this way it provides basis to assess the likely impacts of programmes (The World Bank, 2004).

By mapping out the determining or causal factors judged important for success, and how they might interact, it can then be decided which steps should be monitored as the programme develops. This allows the critical success factors to be identified.

Grounded Theory

The discovery of theory from data is called grounded theory. It provides us with relevant predictions, explanations, interpretations and applications in the area of socio-psychological, extension education and community development problems confronting today. Good theory is produced by combination of an inquiring mind, rich experience and stimulating data. In this direction on many instances, both forms of quantitative and qualitative data are necessary, as supplements, for mutual verification and comparison that will generate theory. Comparative analysis is a strategic method for generating theory from the social units, and is used to generate two basic kinds of theory: substantive and formal.

Substantive and Formal Theory

The substantive theory means that the theory developed as a substantive one or empirical area of sociological inquiry. The formal theory means that the theory developed for a formal or conceptual area of sociological inquiry. Substantive theory is faithful to the empirical situation, which cannot be formulated by applying a few ideas from an established formal theory to the substantive area. The grounded theory approach allows substantive concepts and hypothesis to emerge on its own and enable the analyst to ascertain which, if any, existing formal theories may help him to generate his substantive theories. Substantive theory thus helps to generate new grounded formal theories and to reformulate previously established ones.

Elements of the Theory

Based on grounded theory, the elements of theory are generated by comparative analysis; first, conceptual categories and properties; and second, hypotheses generalizations among the categories and their properties.

(i) Properties and Categories

A property is a conceptual aspect or element of a category. *Both* categories and properties are concepts indicated by the data. The type of concept that is generated has two essential features. First, the concepts should be analytic and sufficiently generalized to designate characteristics of concrete entities, not the entities themselves. Second, the concepts should be sensitizing to yield a "meaningful" picture, abetted by apt illustrations that enable one to grasp the reference in terms of one's own experience. To make concepts both analytic and sensitizing the reader has to see and hear vividly the people in the area under study.

(ii) Formulation of Hypotheses

The comparison of differences and similarities among social groups not only generate categories, but also rather speedily generate generalized relations among them. It must be emphasized that these hypotheses have at first the status of suggested, not tested, relations among categories and their properties, though they are verified as much as possible in the course of research.

In field work, however, general relations are often discovered, as the field worker literally sees them occur. This aspect of the "real life" character of field work deserves emphasis, as it is an important dividend in generating theory. In the beginning, one's hypotheses may seem unrelated, but as categories and properties emerge, development of abstraction becomes related. Their accumulating interrelations form an integrated central theoretical framework.

3. Integration

Integration of the theory takes place at many levels of generality that emerge which necessitates a distinction between "working" and "theoretical" hypotheses. If researcher's emphasis is on integration, then it takes into consideration the fullest range of conceptual levels. Anyone who uses the integrated theory can start at a more general level and focus upon a specific area within the theory, work down to data, still guided by hypotheses for limited, specific situations.

It must be emphasized that integration of the theory is best when it emerges like the concepts. The theory should never just be put together, nor should a formal theory model be applied to it until one is sure, it will fit, and will not force the data. Possible use of a formal model of integration can be determined only after a substantive model has sufficiently emerged.

For substantive theory, the analyst is very likely to discover an integrating scheme within his data since the data and the interrelations of his theory lie so close together. However, the comparative analysis of diverse kinds of substantive groups, though aim at generating "grounded" formal theory, can take the researcher far from emergent substantive integrations. Then existing formal models of process and structure and analysis become useful guides to integrating the categories of a formal theory provided that integration is not forced on the theory. Models of integration for substantive theory that are derived from the data are not necessarily applicable to other substantive areas. Their transfer should be attempted with great caution, and only after trying to discover an emergent integration first.

To conclude, joint collection, coding, and analysis viz., of data underlies operation. The generation of theory, coupled with the notion of theory as process, requires that all three operations, be done together as much as possible. They should blur and intertwine continually, from the beginning of an investigation to its end. The tendency is to do all three simultaneously; but in most of the studies of description and verification, there is typically such a definite focus on one operation at a time that the others are slighted or ignored. This definite separation of each operation hinders generation of theory.

4. Application of Grounded Theory

The practical application of grounded theory, whether substantive or formal, requires developing a theory with four highly interrelated properties. The first requisite property is that the theory must closely fit the substantive area in which it will be used. Second, it must be readily understandable by laymen concerned with this area. Third, it must be sufficiently general to be applicable to a multitude of diverse daily situations within the substantive area and not just to a specific type of situation. Fourth, it must allow the user partial control over the structure and process of daily situations as they change through time. These four requisite properties can be briefly termed as, fitness, understanding, generality and control.

(i) Fitness

Grounded substantive theory, must correspond closely to the data if it is to be applied in daily situations. A grounded theory that is faithful to the everyday realities of a substantive area is one that has been carefully induced from diverse data. Only in this way, the theory will be closely related to what is actually going on in a substantive area and be highly applicable to deal with them.

(ii) Understanding

Grounded substantive theory closely corresponds to the realities of an area, make sense and understandable to the people working in the substantive area. This understanding is crucial, since it is these people who wish to apply the theory themselves. Understanding the theory sharpens their sensitivity to the problems that they face and gives them an image of how they can potentially make matters better, through their own efforts.

(iii) Generality

The sociologist involved in generating theory should be guided by the criteria that the categories should not be as abstract as to lose their sensitizing aspect. It must be clear enough to make one's theory a general guide to multi-conditional, ever-changing situations and also flexible enough to make a wide variety of changing situations understandable. The person who applies the theory will believe and be able to adjust or quickly reformulate a grounded theory. When applying it, one should be able to manage the situational realities that one wishes to improve. The person who applies theory becomes in effect a generator of theory and in this instance the theory is clearly seen as process and an ever-developing entity.

To achieve a theory, that is general enough to be applicable to the total picture, it is more important to accumulate a number of diverse qualitative "facts" on many different situations in the area. This diversity facilitates the development of a theory with both a sufficient number of general concepts relevant to most situations and plausible relations among these categories to account for much of everyday behaviour in those situations. The by-product of such changes, occasioned by the application of grounded theory is a correction of inaccuracies in observation, and reintegration of the correction into the theory. A substantive theory must be induced with its own general concepts, or else a grounded substantive or formal theory that fits the area for application. Then the concepts of these theories become a bridge to more formal sociological categories.

(iv) Control

The substantive theory must enable its users to have enough control in daily situations to make its application worth trying. The users of the theory must be enabled to understand and analyze ongoing situational realities, to produce and predict change in them, and to predict and control consequences both for the objectives of change and for other parts of the total situation that is affected.

As changes occur, the theory must be flexible in revising tactics of application and in revising the theory itself if necessary. To give this kind of control, the theory must provide a sufficient number of general concepts and their plausible interrelations. These concepts must provide the practitioner with situational controls and access to the situation in order to exert the controls. The crux of controllability is the production and control of change through "controllable" variables and "access" variables.

A theory with controllable concepts of sufficient generality fits and is understandable. It gives anyone who wishes to apply these concepts to bring about change a controllable foothold in diverse situations. The properties of controllable variables, further indicates, why grounded theory is suited to practical applications. First, the theory must provide controllable variables with much explanatory power and must "make a big difference' in what is going on in the situation that is to be changed.

The grounded substantive theory, therefore, gives participants a broader guide to what they already tend to do, and perhaps help them to be more effective in doing it. But control usually involves the efforts of two parties ; that is, control of the interaction between two groups by one or both of them. Objects and physical spaces are of strategic importance as variables help to control situations and people's behaviour.

The grounded theory to be used in practice must also include access variables. These are social structural variables that allow, guide, and give persons access either to the controllable variables or to the people who control them. To use a controllable variable one must have a means of access to it. Access variables also indicate how best to enter a situation in order to manage a controllable variable which otherwise unduly disrupt the situation.

5. Prerequisites for Discovering Grounded Theories :

- The researcher has to enter his field in an unprejudiced state of mind "without any preconceived theory that dictates, prior to the research, 'relevancies' in concepts and hypotheses".
- Researcher literally ignores the literature of theory and fact on the area under study, in order to assure that the emergence of categories will not be contaminated by concepts more suited to different areas.
- Grounded theory is derived from data and then illustrated by characteristic example of data.
- Generating a theory from data means that most hypotheses and concepts not only come from data, but are systematically worked out in relation to the data during course of the research.
- Constantly interrelating theoretical (evolving concepts and theories) and empirical data aspects as 'analytic core' serves as a guide for further collection and analysis of the data and for further development of concepts.

Conclusion

In generating a grounded theory, particularly a substantive one, the researcher increases its potential for practical applications by including controllable and access variables if they do not emerge by themselves. Grounded theory generated in this way, will fit, be general enough, and be understandable. One property of an applied grounded theory is that the theory is developed only by professionally trained sociologists, but can be applied by either laymen or sociologists. Further, grounded theory is applicable *in* all situations as well as *to* everybody. Thus people in situations for whom a grounded theory has been generated can be applied in the natural course of daily events.

REFERENCES

Glaser, G.B. and Strauss, A.L., 1999. *The Discovery of Grounded Theory*, published Aldine De Grubter, New York.

The World Bank, 2004. *Monitoring and Evaluation: Some Tools, Methods and Approaches*. The World Bank, Washington D.C., U.S.A.

Index